Andrew Wynter

The Borderlands of Insanity and other Allied Papers

Being Essays from the Quarterly and Edinburgh Reviews

Andrew Wynter

The Borderlands of Insanity and other Allied Papers
Being Essays from the Quarterly and Edinburgh Reviews

ISBN/EAN: 9783337345921

Printed in Europe, USA, Canada, Australia, Japan

Cover: Foto ©berggeist007 / pixelio.de

More available books at **www.hansebooks.com**

THE
BORDERLANDS OF INSANITY

AND

OTHER ALLIED PAPERS.

BY

ANDREW WYNTER, M.D., M.R.C.P. LOND.

Author of "Curiosities of Civilization," &c., &c.

Being Essays from the QUARTERLY AND EDINBURGH REVIEWS.

NEW YORK:
G. P. PUTNAM'S SONS,
FOURTH AVENUE AND TWENTY-THIRD STREET.
1875.

CONTENTS.

The Borderlands of Insanity ... ... *page* 1

Non-Restraint in the Treatment of the Insane ... ... „ 74

The Training of Imbecile Children ... „ 164

Eccentricities of the Mentally Affected ... „ 189

Brain Enigmas ... ... „ 213

Hallucinations and Dreams ... ... „ 257

The Suicidal Act ... ... „ 291

TO THE READER.

IN the following pages—in a series of detached articles, all more or less bearing on the same subject—I have endeavoured to show how fine is the line that separates Sanity from Insanity, and how large is the group of persons dwelling in the neutral territory. But slight occasions are sufficient to take them over the frontier; the necessity of watchfulness, ere the line of demarcation is passed, and an individual finds himself deprived of his civil life, is, therefore, apparent, and needs no apology for being dwelt upon. Even in cases where insanity is declared, or mental weakness is confirmed, I have endeavoured to show that— excepting in the more violent cases where the appliances of an asylum are absolutely necessary—it is an error, especially during the period of convalescence, to crowd patients together in an atmosphere sodden with insanity, where they have to struggle through many adverse circumstances in their progress towards mental convalescence. The true method of cure, in the author's opinion, is to surround the patient with sane minds, and this desideratum can best be found in the family of the physician, where the influence, of the family life of one of the best educated classes of the community, aided by professional tact, is by far the best mental medicine that can be applied to the patient.

Even in chronic and harmless cases of insanity, the support of healthy minds is, the writer thinks, absolutely necessary to keep the mentally afflicted from deteriorating; and we cannot give a better instance of the fact than the success, both psychologically and economically, of the ancient community of Gheel, in Belgium, where, for hundreds of years, there has been an Insane Colony living the social life and joining in the labours of the peasantry with the most marked success. In Kennoway, in Scotland, again, which has not inaptly been called the Scottish Gheel, an Insane Colony has for many years been planted, and has been highly eulogized by the Scottish Commissioners in Lunacy. It is only necessary to add that some of these papers have already appeared in the "Quarterly Review" and other publications, and are now first collected as a humble contribution to psychological medicine by

THE AUTHOR.

CHESTNUT LODGE, CHISWICK,
 May 2nd, 1875.

THE

Borderlands of Insanity.

THAT there is an immense amount of latent brain disease in the community, only awaiting a sufficient exciting cause to make itself patent to the world, there can be no manner of doubt.

In the annual reports from lunatic asylums, we see tables of the causes of the insanity of their inmates, which would lead the public to believe that certain powerful emotions were sufficient to disorganize the material instrument of thought. Thus we find love and religion figuring for a very large proportion of the lunacy of our asylums; whilst a fire, a quarrel with a friend, are set down as the causes which precipitate an individual from a state of sanity to madness.

We do not mean to say that any sound psychologist imagines that these causes are anything more than proximate ones; but the public, and possibly medical men little versed in mental disease, seem to think that a healthy mind can be suddenly dethroned by some specific emotion, just as a healthy body may be suddenly prostrated by fever. There is, in fact, no such thing as sudden insanity, or at least it is of the rarest possible occurrence. Coroners' juries imagine that a person who has committed suicide became insane only at the moment of inserting his neck in the fatal noose; but every one who has studied the human mind must be aware that it is not constituted like a piece of cast iron, which snaps suddenly under the influence of a sudden frost-like emotion.

The grey fabric of the brain, before it gives way, always affords notable signs easily capable of being read by an accomplished psychological physician, of a departure from a state of health.

It happens oftener than we imagine that impending lunacy is known to the individuals themselves before any sign is made to others. There is a terrible stage of consciousness in which, unknown to any other human being, an individual keeps up, as it were, a terrible hand-to-hand conflict with himself, when he is prompted by an inward voice to use disgusting words, which, in his sane moments, he loathes and abhors ; these voices will sometimes suggest ideas which are diametrically opposed to the sober dictates of his conscience. In such conditions of mind, prayers are turned into curses, and the chastest into the most libidinous thoughts.

It does not necessarily follow, because a man is haunted by another and evilly-disposed self, that he has reached the stage of lunacy, if his reason still retains the mastery.

It is said that Bishop Butler waged, for the greater part of his life, a hidden warfare of this kind; and yet no one ever suspected him of unsoundness of mind. It is indeed strange what wayward and erratic turns the mind will take even in robust health. For instance, many of us must have felt the difficulty now and then of suppressing the inclination to cry out in church, or to prevent the rebellious muscles of the face from expressing a smile, on occasions when the utmost gravity of demeanour is called for. Again, we are often haunted by an air of music, or some voice will repeat itself with such obstinacy as to annoy and destroy the mind, and often to prevent sleep. These curious phenomena are not symptomatic of brain disease, but they are singular examples of transient conditions of mind, which, when persistent, are clearly allied to insanity. When, therefore, this persistence does arise, a man may be sure then that he requires the attention of his physician, and that there is some cause at work, which is breeding mischief; and unless he attends to the significant warning, the probability is that disease will take a more serious turn, and that the voices believed to be internal will appear external, and lead the unfortunate sufferer to desperate courses.

Possibly the stage of consciousness is the most terrible of all the conditions of mind which lead the way to insanity. The struggles with the inward fiend, which the reason finds it cannot exorcise, must be far more appalling than a condition of absolute madness, in which, very often, the mental delusions are of a pleasing character.

A patient, writing to Dr. Cheyney, says :— " Such a state as mine you are possibly unacquainted with, notwithstanding all your experience. I am not conscious of the decay or suspension of any of the powers of the mind. I am as well able as ever I was to attend to my business. My family suppose me in health; yet the horrors of a madhouse are staring me in the face. I am a martyr to a species of persecution from within which is becoming intolerable. I am urged to say the most shocking blasphemies; and obscene words are ever on my tongue. Thank God, I have been able to resist; but I often think I must yield at last, and then I shall be disgraced and ruined."

Dr. Wigan gives an account of a worthy clergyman, who was possessed, as it were, in this manner, when he was suffering from over-work or want of rest. At these times, when preaching, there would seem to be placed before his eyes some profane book, which the devil tempted him to read in lieu of a sermon. This was a case where the brain was suffering from a want of duly arterealized blood, as he found that violent exercise with the dumb-bells effectually cast out the fiend which tormented him. Exhaustion of nervous power,—over-work,—is, we believe, the source of mental distress of this nature to a much greater extent than the public apprehend. In this age, when the race is neck to neck, and the struggle for life is ever straining men's minds to the breaking point; when the boy has to go through an examination for a clerkship of a more severe character than was demanded for an University degree of old; when the professional man serves a seven years' apprenticeship to science, and but too often a second seven to starvation, is it to be wondered at that the mental fibre, in cases where there is hereditary taint, becomes weakened, and unable to resist the strain of any great excitement, or further process of exhaustion?

It too often happens that the stage of

consciousness is allowed to progress unperceived, the unfortunate sufferer concealing the agony that is eating into his very soul with the utmost jealousy from the wife of his bosom, and from his dearest friends. We have no doubt in our own minds that innumerable acts which puzzle, and appear totally unaccountable to friends and strangers, are the result of mental conflicts hidden in the depths of the patient's mind. In such cases the demon in possession seizes those very moments in which the enjoyment of other men is found. At the festal board, in the moments of conversation with friends, in the company of ladies, when everything is *couleur de rose*, this conflict will sometimes rage the fiercest, and lead the would-be placid partaker of them to sudden movements, or fits of abstraction, which puzzle and confound those who watch his conduct. And yet, in the great majority of such cases, medicine—and by this term we use the phrase in its largest sense, such as change of scene and air, and rest with proper medicaments—is potent to exorcise the foul fiend and to restore the sufferer to his usual mental health.

The dependence of mind upon body is often proved in the most unmistakable manner in such cases. A single prescription, like the abracadabra of the magician, will convert the man on the verge of insanity to his old serenity of mind. An anecdote is told of Voltaire and an Englishman, which admirably illustrates this position. The conversation between the two happening to turn upon the miseries of life, the *ennui* of the Frenchman and the spleen of the Englishman, both so far agreed that they decided that existence was not worth having, and they determined to commit suicide together on the following morning.

The Englishman punctually arrived, provided with the means of destruction, but the Frenchman appeared to be no longer in the suicidal humour, for on the other proceeding to the execution of their project, Voltaire amusingly interposed:— "*Pardonnez-moi, monsieur, mais mon lavement a très bien opéré ce matin, et cela a changé toutes idées là.*" Feuchlersleben, in his Mental Psychology, has very subtly said that "if we could penetrate into the secret foundations of human events, we should frequently find *the misfortunes of one man caused by the intestines of another*"! This may appear a fantastic proposition on the part of the learned German, but do we not, as men of the world, act upon the knowledge of this fact every day of our lives? Who would be fool enough to ask a man a favour whilst he was waiting for his dinner? The irritation we labour under during these few minutes is clearly attributable to an impoverished condition of the blood; it is, in fact, a fleeting attack of the temper disease, which the late Dr. Marshall Hall has proved is an abiding condition of some persons, particularly among the female sex. How many professional men, wearied all day by press of business, their blood poisoned by sitting for hours in the dark stagnant air of city chambers, will resume their work after dinner, and even prolong it into the night? How many clergymen, ambitious of distinction in the pulpit, will exhaust their brain by the incessant manufacture of bad sermons? Happy the man who retires behind his bandanna, and aids digestion, and refreshes his brain by the legitimate forty winks.

No man after middle age, if he hopes to keep his mind clear, should think of working his brain after dinner, a season which should be given up to enjoyment. The immediate result of post-prandial labour is always inferior to that produced by the vigorous brain of the morning. When mental labour has become a habit, however, we know how weak are the words of warning to make a sufferer desist; and we are reminded of the answer made by Sir Walter Scott to his physicians, who, on his last illness, foresaw that his mind would break down unless he desisted from brain work : " As for bidding me not to work," said he, sadly, " Molly might as well put the kettle on the fire, and then say, '*Now, don't boil.*' "

It must not be supposed, however, that we wish to deprecate even severe mental labour, when the brain in health is accustomed to it; on the contrary, a well-organized brain demands exercise, and like the blacksmith's arm flourishes on it. We believe that pleasurable productive brain-work can be carried on to an almost unlimited extent without injury.

A poet in the full swing of his fancy, a philanthropist working out some scheme for the benefit of humanity, refreshes rather than weakens his brain. It will be found that the great majority of those who have gained high honours in our universities, have also distinguished themselves in after-life. It is the hard thankless task-work which tears and frets the fine fabric of the brain; it is the strain and anxiety which accompanies the working out of great monetary transactions which produces that silent and terrible *ramollessment* which gradually saps the mind of the strong man, and reduces him to the condition of an imbecile.

When we warn the reader to take notice of "early warnings," it matters not whether the symptoms are those which lead to the entire destruction of motive force, and the obliteration of his powers of motion, or whether they take the road to the more fatal derangement of moral and intellectual qualities: if allowed to proceed unchallenged, they lead alike to the destruction of the individual as a free agent. They are equally brain diseases, for the old idea that there is such a thing as derangement of mind without any lesion of the instrument of thought has long been exploded. This idea probably arose from the fact that, in the vast majority of the brains of the insane, when examined after death, there was no appreciable sign of change—nay, the brain has suffered very severe injuries, and yet been followed by no symptoms of mental disturbance. The changes that take place physically are of too delicate a nature for our science to reach in its present condition; but there seems to be no doubt at all, that abnormal mental phenomena depend upon some unhealthy state of the blood. Polished steel is not quicker dimmed by the slightest breath, than is the brain affected by some abnormal condition of the vital fluid. In the horrible phantoms simulating the thoughts of the insane which haunt us in nightmare, we have a familiar example of the manner in which an overloaded stomach will disturb the mind; in the ravings of the insane consequent upon the prolonged drinking of salt water, in cases of shipwreck; in the temporary effect produced upon the temper by waiting for dinner; and finally in the delirium attending fevers and drinking, we have other and equally well-known cases of mental disturbance inevitably following the absorption of some poison into the blood, or of the starving of it, of some of its nutritive constituents.

The more the fact of the physical nature of insanity is acknowledged, the more it is recognized as an ailment, which can be reached by physical agents, the greater will be the chance of its successful treatment. If a man shivers, and feels depressed, he seeks the advice of his physician that he may meet the coming fever with the best resources of his art. If a man feels his brain disturbed—if he feels the " early warnings" of which his friends as yet know nothing—would it not be equally wise of him to summon the aid of medicine before it is too late? Insanity when not clearly hereditary, if taken in its earliest stages, is more easily cured than many diseases which a man passes through without any great fear; for instance, we question if pneumonia is not less curable than a first attack of insanity. If such a mystery were not made of mental disease, it would be deprived of half its terrors, and of half its evil consequences at the same time.

Whilst we should be keenly alive to the first symptoms of a departure from an ordinary state of mind or habit, it must not be supposed that we see a madman in every individual who thinks for himself, or acts in a manner different from his neighbours. We wish to drag no garden-roller, as it were, over character, and to declare that any person who goes out of the general dead level is to be suspected of being what is called "touched." There are naturally crooked sticks as well as straight ones. If, however, a man habitually of an eccentric turn of mind, were to become all at once like other people, and remained so, we should feel certain that some mischief was brewing. *It is the sustained departure from the normal condition of mind and mode of life which should suggest a grave suspicion of impending insanity.* When we find a modest man become boastful and presumptuous, a lover of truth transmuted into an habitual liar, a person of known probity condescending to petty thefts, a humane individual suddenly turned cruel, and a cautious man wild, reckless, and extravagant, then we may be sure there is mental disturbance of a very severe character. The reasoning power may remain clear, and the intellect as bright as ever, and in the course of a long conversation friends may not perceive the slightest cloud on the understanding. Nevertheless, the reader may be certain that these deviations of the moral sentiments are the switch-points which indicate the fact that the mind is leaving the main line, and that, if left to itself, it will speedily career to destruction. It sometimes happens that such changes take place without their being made apparent even to the nearest friends, and that some trivial conversation or circumstance having led to a suspicion of mental unsoundness, upon inquiry it has been discovered that the individual has half-ruined himself. Esquirol mentions a case of this kind, the subject of which was a merchant of considerable position and fortune, whose hidden alienation of mind was brought to light by his having purchased, at a high price, some very inferior pictures; a dispute respecting their value thereupon arising with his children, he flew into a passion, and his insanity became evident. His children, alarmed at his condition, looked into his affairs, when they were found to be utterly in disorder, and full of blanks. This irregularity had existed for six months, and had there been no discussion respecting the pictures, leading to the discovery of the state of his mind, one of the most honourable mercantile houses in France would have been seriously compromised; for a bill of exchange of a considerable amount had become due, and no measures had been taken to provide for its payment.

The latent seeds of insanity very often become known to the world through unusual physical signs: muscular agitation succeeds to the ordinary repose of the individual. The man who in a state of health is grave and gentle, suddenly puts on a brusquery which astonishes his friends. It would seem as though he sought to stifle his agonizing thoughts by the exhaustion of his physical strength. Incessant change of motion, and an irresistible restlessness give note of the condition of his mind, which finds no rest in any direction, and in the end some maniacal outburst is pretty sure to take place.

But these are the more prominent warnings of coming trouble which cannot well be overlooked. The symptoms we more particularly wish to draw attention to, are those slight deviations from a normal condition which are rarely observed by the sufferers themselves or their friends. One of the most constant and characteristic is a debilitated power of attention. Possibly the most comprehensive definition of genius, is the power of concentrating and prolonging the attention upon any one given point. It is the quality of mind which raises one man above another, and it is the parent of all creations and of most discoveries; and we may add, it is the morbid excess and indulgence of this quality which leads sometimes to mental disease : hence the common observation that genius and madness are only divided by a very thin partition. " The difference," says Sir William Hamilton, " between an ordinary mind, and the mind of Sir Isaac Newton, consists principally in this,—that the one is capable of a more continuous application than the other ; that a Newton is capable, without fatigue, to connect inference with inference, in one long series, to a determined end ; while the man of inferior capacity is obliged to break, or let fall the thread he has begun to spin." This is in fact what Sir Isaac Newton with equal modesty and shrewdness himself admitted. To one who complimented him on his genius, he replied, that if he had made any discoveries it was due more to patient attention than to any other quality. There is, however, a certain morbid attention, when directed towards supposed ailments of the body and mind, which is to be especially deprecated. A man may so concentrate his attention upon certain organs of the body as to create disease in them. The hypochondriac, for instance, never ceases to dwell upon the condition of his digestive organs, and the consequence in the end is that he directs so much nervous energy to the spot as to cause congestion, and actual disease. We see no reason to doubt that mere disordered functions of the brain may be converted by the same undue attention into positive disorganization and mental disease. Hence overstudiousness on these points is to be avoided.

In the majority of cases there is no danger of such a result, but in persons of a highly nervous temperament it is different, and with them the very first step towards health, would be to enable them to get rid of themselves.

Of a similar nature to this exaltation of the faculty of attention, is the exaggeration which often takes place in the special functions of sense. The approach of brain disease is often heralded by the most marvellous exaltation of sight, smell, taste, and hearing. Dr. Elliotson mentions a patient who, previous to an attack of Hemiplegia, felt such an extraordinary acuteness of hearing, that he heard the least sound at the bottom of his house. His vision was also exaggerated to that degree, that he could tell the hour by a watch placed on a table at such a distance as would in a state of health have precluded his even distinguishing the hands.

The sense of smell is often in cases of disease equally increased in force, and often so completely altered, that slight perfumes will be perverted into the most disgusting smells.

In this condition of brain, the avenues by which the outward world is brought in connection with the inward man, are thrown open so wide that it would seem as though the unhappy person projected his special organs of sense outward, until they absolutely came in contact with the objects or manifestations submitted to them. A more distressing condition it would be difficult to imagine, or one which more clearly points to an inflammatory condition of the brain.

The sense of touch is often perverted in incipient brain disease; some invalids assert that everything they touch is like velvet. Andrial noticed that six weeks before a paralytic attack, a patient complained of one half of his scalp feeling like a piece of leather. In the case of a gentleman who died of apoplexy there was for some time previously to his illness, a feeling in both hands as if the skin was covered with minute and irritating particles of dust or sand. He repeatedly complained of this symptom, and was frequently observed to wash his hands as if to remove the imaginary particles. In another case, some time previous to a paralytic seizure, the patient imagined that he had extraneous particles of dirt and stones in his boots, or inside his stockings, irritating his feet and interfering with his personal comfort, as well as his powers of locomotion. This perverted state of sensation was observed for two months previously to an attack of acute cerebral disorder.

A patient under my own care suffering under symptoms of brain disease, fancied everything he touched was covered with grease, to get rid of which he was incessantly washing his hands; indeed, his whole body, according to his own account, was contaminated in the same manner, and in order to cleanse himself he was perpetually taking baths. In fact, he lived in cold water, and yet always protested that he was as greasy as ever. Ultimately, this gentleman was obliged to give up the appointment he held, and is now totally incapacitated for any brain work.

To those unaccustomed to read the subtle indications by which the failing brain gives its warnings, these trifles, light as air, may seem to be of too trivial a nature to warrant the interposition of medicine, and those who venture to draw attention to them are liable to ridicule. On the occasion of the discussion of the Lunacy Amendment Act, not long since, the then Lord Chancellor remarked the tendency of medical men to intrude their "theories" respecting insanity, when acting as witnesses in the law courts. In confirmation of this opinion he read from Drs. Bucknell and Tuke's "Psychological Medicine," a passage which spoke of "a shrivelled ear and bristling hair" as being symptomatic of a certain mental condition. Now, curiously enough, this "shrivelled ear" and "bristling hair," which his lordship laughed at so immoderately, is a most undoubted sign of chronic dementia.

The premonitions of epileptic attacks are too well known to require attention at our hands; and they are, at the same time, so varying in their character, as to preclude reliance upon any one warning symptom. Herein the patient must minister to himself. But the community is not aware that epileptic attacks may go on for years without discovering themselves either to the individual, or to his friends or medical man. In children especially, attacks sometimes come on in the night, and pass away without leaving any sign. Dr. Marshall Hall has done lasting service by drawing the attention of the public to this obscure form of a well-known disease, and the nursery is thus supplied with a hint of great use to the rising generation. These hidden seizures, however, sometimes take place in after-life, and the slightly-bitten tongue, often so slightly indented that it is scarcely perceptible, is the only indication that a symptom of approaching brain disease, of a severe type, has visited the individual in his sleep. Strange as it may appear, however, the most marked and terrible seizures are sometimes mistaken by persons suffering them for the visitations of preternatural agents. Dr. Gregory, of Edinburgh, used to give a case of this kind in his lectures, which is so curious that we shall here relate it. One of his patients told him he was in the habit of dining every day at six, but that he was plagued with a visitor at that hour who always greatly distressed him. Exactly as the hour struck the door opened, and an old hag entered, with a frowning countenance, and with every demonstration of spite and hate, rushed upon him and struck him a severe blow upon the head, which caused him to swoon for a time of shorter or longer duration. This apparition, he asserted, was of daily occurrence. Dr. Gregory, guessing that some mental delusion was at the bottom of this singular attack, invited himself to dinner with his friend, adding, "We will see if your malignant old woman will venture to join our company." The gentleman gladly accepted the proposal, expecting the doctor's ridicule rather than his sympathy. When the dinner arrived, the doctor exerted his powers of conversation, which were of a very brilliant character, in the hope of diverting his friend's attention from the thoughts of the approaching visit, supposing that he was suffering from some nervous attack, and he so far succeeded that the hour of six came almost unnoticed, and he was hoping that the dinner would pass without the unwelcome interruption. The clock had scarcely struck, however, when the gentleman exclaimed suddenly, in an alarmed voice, " The hag comes again," and dropped back in his chair in a swoon, in the way he described. These periodical attacks were clearly traced to sudden head seizures, which gave way to the appropriate remedies.

Whilst an exaltation of the faculty of attention points to insanity, the growing deficiency of it points as certainly to a coming imbecility, and especially of an impending attack of softening of the brain— that terrible affliction, which may be termed the stockbrokers' disease, so liable do the *habitués* of Capel Court seem to its visitations. The first beginning of the disease very often comes upon a man in the height of his prosperity, and its approach is so insidious that although he may be walking about and transacting his business, this fatal rot may have already commenced. As in the "Vision of Mirza," a passenger is every now and then missed from the ever-ebbing and flowing stream of life, and none but the physician notes that he has dropped through the pitfall on the bridge, and will never mix in the busy haunts of man again.

In the early stages of cerebral softening, when the delicate nerve-vesicle begins to disintegrate, a debility of memory becomes apparent. The most common affairs of life are forgotten; names, dates, pass out of recollection, and whole passages of the patient's existence fade away from his life, as it were. Even whilst he is engaged in some occupation which requires a recollection of what he has done before, he finds he is wholly at fault; matters he has been informed upon over and over again are entirely forgotten. At such times the character of the handwriting affords a singularly accurate gauge of the deterioration the mind is undergoing.

Handwriting echoes, as it were, both the physical and mental condition of the calligraphist. The very attitude of his nerves, so to speak, is indelibly registered in the manner in which he directs his pen; and no better method of testing the departure of the patient from his normal state of health can be fixed, than by comparing his handwriting at different stages of the progress of his disease. The erasures that take place in the letters of the patient as the malady progresses are very noteworthy; the misspelling, the hesitating efforts at expressing his ideas, mirror, as it were, the confusion that reigns in his mind.

We are inclined to think that the sign of cerebral softening most to be dreaded is the want of power to fix the attention. A person may suffer from temporary loss of memory from very slight causes; such, for example, as exhaustion. Sir Henry Holland, in his very interesting " Mental Pathology," tells us that, having descended two deep mines in the Hartz Mountains, and having undergone much exertion without food, he found himself suddenly deprived for a short time of his memory, which returned again immediately after taking food and wine. A copious draught of wine will often restore these momentary fits of loss of memory, which are dependent upon no organic disease, but arise from want of proper circulation in the brain. We all know, when we have forgotten a particular name or thing, the pertinacity with which it seems to recede further from the memory, the more we try to recall it to mind; it remains upon the tip of the tongue, but will not come forth. These are familiar examples of transient loss of memory, which only prove how often the healthiest brain is for a moment plagued with symptoms of no account, but which, when persistent, are the invariable precursors of serious brain disease.

There are certain significant, although but slightly-marked signs of softening, which tell clearly to the eye of the practised physician the approach of the disintegration of the cerebral matter. The trained eye will observe a loss of muscular power; the patient will slip on one side; the leg is put forward with great premeditation; volition ceases to act unconsciously; certain acts are performed as though the sufferer were pulling the wires of a doll; the hand cannot grasp with a firm, healthy grip. A minute degree of facial paralysis will sometimes disturb the wonted expression of the countenance, without even friends knowing the cause. A very slight elevation of one eyebrow, a drawing aside of the mouth a hair's breadth, will materially alter the look of a person ; and paralysis of this kind often exists without any one suspecting that softening of the brain is impending. This partial paralysis, which is indicative of approaching apoplexy, very often shows itself in a person's speech. When we remember the number of muscles which must co-ordinate to enable a man to articulate, it will be readily understood that any loss of power in these delicate muscles must show itself in the speech. It often happens that the first signs will be a clipping of the queen's English : the person will speak as though he were drunk ; indeed, drunkenness does produce the very temporary paralysis we allude to.

A still more singular sign of softening, and the apoplexy that results, is the odd way in which persons in this position will transpose their words. Dr. Beddoes mentions the case of a gentleman who, previous to an attack of brain disease, used to commit laughable blunders of this kind. For instance, he would say, " Everybody feels very languid this *wet* weather—I mean this *hot* weather ;" or, " Come, who will sit down to supper ? Here is cold meat and *pudding*—I mean *pie.*" Undiscovered and partial paralysis is the cause sometimes of odd mistakes. Thus a gentleman angrily demanded of his servant whilst at dinner why he had brought him a broken wine-glass. The servant, on examining it, affirmed it was a sound one. The master again scolded him, but on inspecting it himself, found it to be really unbroken. The explanation of this circumstance was that the gentleman had suddenly been seized with paralysis of the nerves of sensation of one side of his lip; consequently, as there was no feeling there, within a certain circumscribed space, he naturally concluded, without looking, that a piece of glass had been broken away. In other cases, a person will declare that his fingers feel like a sausage. Early warnings these, if persisted in, that should not be neglected for a single moment.

The sight also gives warnings, that are equally unmistakable to the physician of coming trouble, and more especially the dread symptom of double vision. Dr. Gregory tells a curious and highly instructive tale of a sportsman, who, when out shooting one day with his gamekeeper, complained of his bringing out so many dogs, asking why he required eight dogs. The servant said there were only four, but his master persisted that there were double that number. Convinced, however, of his mistake, probably by the touch, he immediately became aware of his condition, mounted his horse, and rode home; and had not long been there before he was attacked with apoplexy and died. It must be remembered, however, that double vision may arise from a far more innocent cause than brain disease. Mr. Brudnell Carter has only lately pointed out that it may arise, with a consequent vertigo, from a very short sight, compelling a reader to hold his book close to the eyes, and the convergence thereby induced, resulting in double vision. Indeed he gives a case where the symptom, showing itself in an Oxford student studying for honours was so afflicted, and the symptoms so alarming his medical man, that he was induced to leave his University without taking a degree, and to go to Australia for the benefit of his health; from which place, however, he returned without improvement. In this condition he applied to Mr. Carter, who cured him by directing him to use spectacles, and to practise reading at a distance of eighteen inches from his eyes.

It is not very easy to distinguish softening of the brain from another malady, which is equally terrible. We allude to the general paralysis of the insane. Indeed, the latter disease is very often but a result of the other. General paralysis of the insane may indeed be impending for years, the only symptom being an exaggeration, as it were, of an ordinary mental failing. The individual may be seen gradually to become very exaggerated and extravagant in his ideas. If in business, all his speculations, he will assert, have turned out beyond his hopes. His prospects are growing more brilliant every day, and he will launch out in extravagant modes of living, often to the surprise of his friends, who have been wont to look upon him as a careful, considerate man. This mental transformation will often exhibit itself for years without any further symptom showing itself. The next symptom that shows itself will be, perhaps, a slow and measured method of intonation. Phrases are selected with the utmost care, the lips sometimes opening and shutting without giving out a sound, assuming, in the action, the smoking of a pipe. Indeed, the mouth opens and shuts in one piece, neither lips or the muscles of the mouth yielding the least expression. At this stage the mental and physical powers of the patient are perceived to gradually give way; all his muscles become gradually paralyzed. Even those reflex actions which preside over so many of the functions of the body die; and it often happens that a patient is suddenly choked by the passage of the food down the windpipe, instead of the gullet,— the epiglottis, that sensitive lid which, in a state of health, so jealously closes and guards the air-passage, being paralyzed, and standing open, as it were, to invite the dissolution of the body, thus reduced to a living death.

The injurious effects of blows upon the head are not sufficiently considered, for the reason that in many cases they don't show themselves for years. Where any serious concussions of this kind have taken place, the individual receiving them should always beware of the first signs of distress in the brain. Numberless cases are on record in which a fatal termination has ensued from a blow on the head, received years previously. A sailor fell from the mainyard of a ship upon deck, and was removed below in a state of unconsciousness. He speedily recovered his senses, however, and in a fortnight resumed his work. No bad symptoms occurred for four years, after which he was occasionally attacked with headache, and twenty-six years afterwards he became paralytic, in which state he continued for eight weeks, when he died, and on examination it was discoverd that a large abscess existed in his brain. In another case a boy received a violent blow from a cricket-bat, from which he did not suffer any inconvenience for ten or eleven years, when he became liable to attacks of headache of a severe character; epileptic seizures followed, and he ultimately died, when an encysted abscess, of the size of an egg, was found in the cerebrum.

On the other hand, there are some remarkable examples on record of individuals who have been suddenly seized with disease of the brain, who after a time have been suddenly restored to consciousness, and have at once reverted to the action in which they were engaged. During the battle of the Nile a captain was struck on the head whilst in the act of giving orders. A portion of the skull was driven in upon the brain and the officer at once became unconscious. In this condition he was taken home, and removed to Greenwich Hospital, where he remained for fifteen months, living the life of an inanimate vegetable; upon the operation of trepanning being performed, however, his consciousness immediately returned, he rose up in bed, and in a loud voice finished giving the order he was issuing when he was struck down.

The hereditary nature of insanity has long been acknowledged by the physician and by society in general, but, with the exception of those who have specially applied themselves to this branch of medico-psychology, the bare admission of the fact is all that is known about one of the most interesting subjects in the realm of medicine. The general idea is that madness begets madness—that the children of insane parents are either free from the disease or inherit the calamity in all its integrity. All of us who have had any practical experience in the treatment of insanity, however, know full well that nature never transmits her pattern with exact fidelity. The likeness of the child to its parents, how like and yet how unlike. In a still more marked degree is the mental likeness in the majority of instances transmitted with a difference, although in kind. These analogies are necessary to show that the same rule obtains in the transmission of mental disease. It is but rarely that the same form of insanity is passed on from parent to child. In many cases, indeed, there is no direct mental disease, such as can be legally taken notice of, but all the children, more or less, by the virtue of their inheritance, have passed into the Borderland of insanity. According to the intensity of the inheritance, will be the degree of the mental perversity transmitted. It may assume either a bodily form, as in the shape of neuralgia, epilepsy, chorea, or it may show itself in mental peculiarities, many forms of which may seem liberally exaggerated instances of moral failings; such as violent passion, acts of cruelty, thieving, lying, &c.

When we remember the number of persons in the country whose insanity is undoubted, it will be admitted that there must be a very large number of individuals who inherit either the disease direct, or are saturated with the seeds of nervous disorders, which only require some exciting cause to force them into vigorous growth. It is this class of incipient lunatics with whom we wish to deal in these following pages—persons whose unsoundness of mind is mistaken by the world either for mere eccentricity or for moral perversity. Going about the world with a clean mental bill of health, as it were, these unfortunate individuals are the cause of more misery to themselves and to the world than the proclaimed lunatic, for the reason that society cannot restrain them from acts as destructive to the well-being of others as to themselves.

One of the ablest alienist physicians of the day, Dr. Maudsley, writing of this unhappy class, says:

"It certainly cannot be disputed that when nothing abnormal whatever may be discovered in the brains of persons who have a strong hereditary tendency to insanity, they often exhibit characteristic peculiarities in their manner of thought, feeling, and conduct, carrying in their physiognomy, bodily habit, and mental disposition, the sure marks of their evil heritage. These marks are, I believe, the outward and visible signs of an inward and invisible peculiarity of cerebral organization. Here indeed, we broach a most important inquiry, which has only lately attracted attention; the inquiry, namely, into the physical and mental signs of the degeneracy of the human kind. I don't mean to assert that all persons whose parents or blood relatives have suffered from nervous or mental disease, exhibit mental or bodily peculiarities: some may be well formed bodily, and of superior natural intelligence, the hereditary disposition in them not having assumed the character of deterioration of race; but it admits of no dispute that there is what may be called an *insane temperament* or *neurosis*, and that it is marked by peculiarities of mental and bodily conformation. . . . . What are the bodily and mental marks of the insane temperament? That there are such is most certain; for although the varieties of this temperament cannot yet be described with any precision, no one who accustoms himself to observe closely, will fail to be able to say positively, in many instances, whether an insane person, and even a sane person in some instances, comes of an insane family or not. An irregular and unsymmetrical conformation of the head, a want of regularity and harmony of the features, and, as Morel holds, malformation of the external ear, are sometimes observed. Convulsions are apt to occur in early life ; and there are antics, grimaces, or other spasmodic movements of muscles of face, eye lids, or lips afterwards. Stammering and defects of pronunciation are also sometimes signs of the neurosis. In other cases there are peculiarities of the eyes, which, though they may be full and prominent, have a vacillating movement, and a vacantly abstracted, or half-fearful, half-suspicious and distrustful look. There may indeed be something in the eye wonderfully suggestive of the look of an animal. The walk and manner are uncertain, and though not easily described in words, may be distinctly peculiar.

With these bodily traits are associated peculiarities of thought, feeling, and conduct. Without being insane, a person who has the insane neurosis strongly marked, is thought to be strange, queer, and not like other persons."

Cannot the reader, passing his memory over children who have inherited the insane temperament, recognize in this minute yet truthful sketch many of the physical peculiarities here hinted at? How many children has he not seen with all the signs of that "queerness" which means so much more than we are able to convey in more precise terms. In how many children of this class do we not notice waywardness, extreme perversity of character, great cruelty, and a total want of any moral sense.

Among the more special forms of moral perversity, or, as the alienal physicians would say, insanity, which are transmitted by an insane parent, forms which may not arise for two generations, may be mentioned Kleptomania, or petty thieving; Dipsomania, or thirst-madness; and Pyromania, or incendiary madness. The unprofessional world, especially the lawyers, cannot understand that what they consider to be certain forms of vice should be explained away and removed from the denunciations of the moralists and the punishment of the law by a new-fangled theory of "mad doctors," which they ascribe to undoubted brain disease. That an individual should in all other matters appear to be of sound mind, but that at certain seasons he should be seized with an irrepressible desire to commit theft, arson, or to reduce himself below the level of a beast by means of drink, seems to the unprofessional understanding quite incomprehensible; and the common view—taking this bare aspect of the case—is the right one. There is indeed no such thing as simple thirst-madness, or fire-madness, or thieving madness, or homicidal madness. Those who have watched such cases with professional knowledge and experience, observe that the whole moral tone of the individuals so afflicted is, so to speak, below par. They suffer from a paralysis of the moral sense; invariably they are untruthful, very commonly full of impure thoughts, and always eccentric both in thought and action. They have long belonged to the Borderland of Insanity, in the opinion of those who know them best; but it is only the last supreme act which, in the eyes of the world, takes them over the frontier into the domain of the insane. There are thousands who, lacking the opportunity or the power of will, never indeed do cross the frontier, but remain and swell the vast army of undiscovered lunatics which leavens unsuspectedly the sane population.

It may be as well, in view of a more ready

discrimination of the probability of the symptoms of a person being those of incipient insanity, to remember what experience teaches us as to the relative powers of the father and mother to transmit the insane neurosis to their children. It is agreed by all alienist physicians, that girls are far more likely to inherit insanity from their mothers than from the other parent, and that the same rule obtains as regards the sons. The tendency of the mother to transmit her mental disease is, however, in all cases stronger than the father's ; some physicians have, indeed, insisted that it is twice as strong. In judging of the chances of an individual inheriting mental disease, or, indeed, of the insane temperament, it may not be unadvisable to study the general likeness and character. If the daughter of an insane mother very much resembles her in feature and in temperament, the chances are that she is more likely to inherit the disease than other daughters who are not so like. And the reason is obvious; for if the general physical aspect and the temperament are alike, it points to a smilar likeness in the structure of the body and nerve. The mental likeness is, however, the most important of the two, as we often see children partaking of the father's features and of the mother's temperament. In such cases the child would possibly inherit the mother's insane temperament, transmuted into some disorder of the nervous system, such as hysteria, epilepsy, or neuralgia; for nothing is more common than to find mere nervous disorders changed, by transmission from parent to child, into mental disorders, and *vice versâ*. In short, all later discoveries teach us that they are interchangeable, in some mysterious way, in which our knowledge of the laws of inheritance are, as yet, but dim and faulty. It is also well to know that what may be termed peculiarities, oddities, and all the different small mental and bodily peculiarities, which only amount to what may be termed "queerness," &c., are more generally transmitted by insane parents to their children, than are the more recognized forms of the disease itself.

The most common and, as regards society and themselves, the most terrible of these minor offshoots of the insane diathesis, are the moral diseases, such as dipsomania, or drink-madness. To the ordinary observer the dipsomaniac is nothing more than an utterly reckless person, who is determined to obtain drink, regardless of consequences. He is confounded with the ordinary drunkard, and his infirmity is looked upon as a simple vice. But, in reality, the two cases are utterly unlike. Whilst in the case of the ordinary toper drink is only the accompaniment of the festive board, in the dipsomaniac it is a secret vice. He will, indeed, avoid drinking in company, and assume the virtue of temperance, all the time that he is madly looking for liquor; and when he cannot obtain it, will drink even "shoeblacking and turpentine, hair-wash, or anything stimulating," says Dr. Skie. There is one feature in the dipsomaniac which is very observable; he is invariably good-tempered when not suffering from the physical depression which follows the indulgence of his desire. My own experience of cases under my charge, and which I have watched narrowly, lead me to the conclusion that the dipsomaniac is, without exception, a happy-go-lucky sort of person, with whom the world appears to go smoothly. The worst feature of the disease is the very small percentage of cures which are obtained. Among women, there appears to be more chance than among men, as the irresistible desire, in some cases, leaves them after a certain period of life. But their case is rendered the more distressing, as it usually happens that the most refined natures, under such circumstances, are transformed into the lewdest and the most shameless of their sex.

When the attack is over, the patient is overwhelmed with remorse at the disgrace he has brought upon himself; and this remorse and swinish bestiality alternate until every worldly prospect is ruined, and the poor patient dies in a fit of delirium or is transferred as "a boarder" to the custody of an asylum. But the mere treatment of an asylum, which is a simple withholding of the liquor during the time the irresistible impulse is upon the patient, is good only whilst the impulse lasts; when the fit is over, no asylum proprietor would be justified in retaining his patient an hour. It is true it has been proposed by the select committee on the Habitual Drunkards Bill that the asylum proprietor should retain the patient during the period of the remission of the attack, as well as during its accession; but we do not believe the legislature will ever consent to agree to such an infringement upon the liberty of the subject, which in the case of a confirmed dipsomaniac would be simply imprisonment for life. Neither is an asylum the proper place of seclusion at such times of remission. If the dipsomaniac during the subsidence of the attack is to be strengthened against future impulses in the same direction, it must be by associating with sound minds. This, in my opinion, is an unanswerable objection to the proposition to treat them as boarders in asylums on parole. If the dipsomaniac is to be held only by his honour, there is no need of the asylum walls; he is far better treated in the house of a medical man, where far more individual attention could be given to his case, and where his own efforts at reformation would be strengthened by the sound surrounding minds, and where the irritating signs and emblems of coercion were altogether absent.

This disease being in nearly every case inherited from some insane parent,—although so remote that he may have been forgotten,—in most cases it seems almost useless to appeal to the determination of the patient to resist it. In some cases, however, where the will is stronger than the craving, even moral training is often of great service. We must, however, protest in the strongest language we can use against the modern tendency to put the enemy in the mouths of the individuals prepared by nature as it were to the attacks of the disease.

Not many years since a leading London physician became inspired, almost to fanaticism, with the use of alcohol as a medicine. He prescribed it in varying doses in nearly all diseases; even fevers were not exempt; and the teaching of this master has been handed down to a certain number of disciples, who are but too willing to improve upon his teaching. This unfortunate initiative, under the guise of science, has in a large number of cases led to confirmed drinking on the part of ladies; and of course all those predisposed by a bad inheritance have been led into the trap held out to them by those who should have been the first to lead them away from it.

We all know the ease with which hysterical depression makes but too many seek what they euphemistically term "support" under such circumstances.

Beginning with sal-volatile, and ascending by graduated steps, such as red lavender, eau de Cologne, to sherry and spirits, it may easily be conceived that the liberal prescription of brandy by a celebrated physician, as one of the most efficacious and universal of remedies in diseases (usual amongst them) was calculated but too rapidly to be accepted and acted upon. Hence the necessity for the reasoning that " alcohol in whatever form should be prescribed with as much care as any powerful drug; and that the sanction of its use should be so framed as not to be interpreted as a sanction for excess, or necessarily for the continuance of its use when the occasion is past."

But let us ask, is there no other cause of this female intemperance besides the injurious misuse of the physician's prescriptions? Have the changes which have taken place in social life anything to answer for?

Within these last twenty years the railway may be said to have driven female society farther and farther into the country. The man goes forth to his labour in the morning and returns in the evening, leaving his wife, during the whole of the day, to her own devices. Whilst our residences were in town, they always had that intensely feminine refreshment, shopping, to solace their *ennui;* but shops now being out of the question, what have our wives to do, especially the childless ones, under the present miserable views as regards their education? It has been, we think, well said that the working man goes to the public house for light and cheerful society as much as for beer. After the muscular fatigues of the day, he requires some social recruitment of this kind. May we not ask if some recruitment for the woman's mind is not required after the humdrum housekeeping labours for the day are accomplished? We think this cannot be denied. As it is, she is left to her own devices, and if there is no family to be attended to, we know that these devices are not of too intellectual an order. In short, arrangements of society and the railways have banished our wives from all the amusements and excitement of town, and they are thrown upon their own resources with but very imperfectly educated minds. The result is, in many cases, where this incipient form of insanity is present, a fatal appeal to the bottle. As long as ladies are ashamed to put their hands to any useful matter, and are untrained to any intellectual work, they cannot help getting into mischief. For this reason we believe that the great cure for the evil that has begun to show itself within the sacred precincts of home, is a good intellectual training for women. If our wives knew anything of art, if they could draw, paint, model, or write, we should hear far less of the sherry-bottle. The few hours that elapsed before the husband's return from business would be bridged over by some occupation that delighted and satisfied the mind ; at least the demon that, unknown to themselves, exists within themselves, may not be called into life.

It certainly is deplorable to think that the delicate sense of woman for beauty, her appreciative faculties, her delicate touch, and mental fibre should be wholly lost on the world's work. All that man wants intellectually she is fitted by nature to supply; but unfortunately Mrs. Grundy stands resolutely in the way, and it is thought beneath her position and dignity to do anything useful or ornamental in life, especially if she is paid for it. As long as we bring up young girls with these absurd and snobbish notions how can we expect that anything but evil can come of it? If they are neither allowed to think nor act out of the rigid line grooved for them by what is termed "good society," what right have we to expect that they will become sensible wives, or that they will resist the temptation to drown in intemperance the mental vacuity to which they are trained from their cradle, when we find man, with his thousand occupations, but too often fall into the same pit? Were it not that it is clear women are beginning to rebel against the barriers that are placed against their taking their proper place in society as the helpmate, and in a certain sense the equal of man, we should indeed fear that the habit of intemperance which once was so fashionable in even good society amongst "gentlemen," would root itself amongst what some are pleased to term the softer sex; and if, unhappily, it should do so, the blame would rest mainly with society for having brought about the mischief.

What we have said with respect to the absurdity of the teaching that the dipsomaniac is mad only on the point of drink, may be also said of the kleptomaniac, or individual seized with the thieving mania. The law hitherto has only looked askance at the doctrine that a man or woman can be afflicted by an irresistible impulsive madness to break the eighth commandment only, whilst they are sane and obedient with respect to all the others. The struggle that is still going on in the courts of law between the medical advocate and the barrister, touching the fact as to whether a theft committed by a person afflicted with this mental weakness, is to be considered as a vice or a disease, would we believe cease, if the physician were to assert, as he doubtless would do if left to himself, that, like dipsomania, it is only one of many symptoms of moral insanity. The reason, no doubt, why he does not so insist, is that in such cases the friends of the person charged with theft do not wish the plea of insanity to be urged, for the sake of the family, no less than for the sake of the so-called culprit; as it would, in case of conviction, result in his incarceration in one of the criminal lunatic asylums—a worse fate, in the friends' view of the case, than a mere committal to prison for an ordinary theft. The kleptomaniacs that we have had under our own care, have without exception shown signs of being perfectly oblivious to any moral law. They have been untruthful to the last degree, inventing needless lies; without any feeling, vain, passionate, thoughtless,—in short it would seem as though they were lacking in every virtue that should adorn the honest woman or man. The mere fact of their committing a theft was only one sign that they were irresponsible for any act they may do in contravention of the standard morality which keeps society select. But the differences on this head are, we think, likely to be dissipated by the fact that tradesmen in the majority of cases in the West-end have accepted the position that the dipsomaniac is not a thief in the ordinary acceptation of the word. There are several high families whose weakness in this particular is well known. Their thefts are noted by the shopkeepers, and an account of their depredations is simply sent home, and paid for without further demur, in cases where the articles are not returned by the friends of the unhappy individual, immediately they are discovered.

Among other curious examples of the difference between the ordinary thief and the kleptomaniac, we may adduce the singular evidence of Dr. A. Peddie, an Edinburgh physician, who, in his evidence given before the select committee on habitual drunkards, says, of the style of theft, under conditions of drunkenness, that they were systematic and peculiar. " The sheriff gave an instance of a man who when drunk stole nothing but Bibles; and he was transported for the seventh act of Bible stealing. Then another man stole nothing but spades; a woman stole nothing but shoes; another stole nothing but shawls; and there is a curious case, the indictment of which the sheriff sent me, against a man of the name of Grubb, who was transported for the seventh act of stealing a tub; there was nothing in his line of life and nothing in his prospects, no motive, to make him especially desire tubs, but so it was, that when he stole it was always—except on one occasion—a tub." These examples prove that dipsomania and kleptomania are co-existent in some diseased minds, and that both forms of disease are altogether different from the ordinary vices, which they simulate.

It is a bold but not altogether unwarranted assumption that the many forms of moral insanity which we find presenting themselves, in the upper classes, in the form of the drinking and thieving mania, are due to the evil inheritance their forefathers have transmitted to them from the roistering ages of the Georges, when sobriety even in the best classes was the exception rather than the rule.

The temperance and teetotal folks are not aware of the powerful weapon they have in their hands in the known fact that persistent drunkards, in nine cases out of ten, plant the seeds of insanity and the allied nervous diseases in their offspring—Dr. Howe, of the Idiot Asylum of Massachusetts, says that out of 300 idiots, 145 had drunken parents ! ! ! Once planted there, the fruits, as different generations arise, may be singularly diverse : whilst in one child there may be merely persistent neuralgia, in another the ancestral drunkenness may assume the form of dipsomania, whilst a third may be affected with irrepressible desire to pilfer, whilst a fourth may be only afflicted with a partial paralysis, or with epilepsy. And any of these children may plant in their offspring direct forms of insanity or idiocy. In the persistent abuse of alcohol, in short, we trace, without the smallest doubt, the planting of the germs of mental as well as bodily disease in the blood, and we do not doubt that it is the cause of a very large percentage of the lunacy in the country. If Sir Wilfrid Lawson could carry his Bill, and if public opinion should so far go with him as to abolish persistent drinking in private, we do not doubt but that the most powerful germ of insanity and its allied diseases would be destroyed in the land. It may be a singular prescription for a physician to depend upon, but we believe that education will gradually destroy in the working and the lower middle classes, the vice of drunkenness, as within the memory of many a living man it has already been destroyed by the action of society in the upper classes. One has only to read the memoirs of great men, statesmen and others, in the reigns of the later Georges, to satisfy ourselves that this is the case; and that if any gentlemen were to forget themselves as our grandfathers then did persistently at the dinner-table, they would be scouted from society. But if this crime was not only tolerated, but approved of by the society of the day, Nature has not forgotten the offence against her laws; and what but an age ago was a mere custom has in many families become a permanent affliction, which may take many generations ere the mixture of fresh and healthy blood shall wipe it out. Let us trust that the warning voice, which all alienist physicians have raised with reference to this vice, may be listened to by the sherry-drinking ladies of this generation, lest the tippling inheritance so many of them have more or less received, unknown to themselves, be strengthened, and transformed into a more permanent and terrible form of insanity in their posterity. It is true that Nature, wearied, as it were, by repeated offences against her laws, sometimes—out of mercy to the race— takes the matter into her own hands in a very summary manner by extinguishing the posterity of the habitual drunkard. There is nothing more clearly ascertained in psychological medicine, than that children conceived in conditions of drunkenness of either parents, are liable to become idiots unable to prolong their race. There is not a physician of experience in these low forms of moral insanity, brought about by persistent drinking, who cannot point to the downfall in one generation of the most intellectual parents to the most abject offspring not possessing any claim to humanity. A more terrible example of the swift manufacture of a perfectly waste material out of what might have been an honour to humanity, the mind cannot conceive. In old Rome the bestial vices of the slaves used to be paraded as an example before the children of their masters, as a warning and terror to them. Oh that dipsomaniacs could be so utilized! But unfortunately, as regards their children, the tyranny of their sad inheritance would only render the example a mockery, a delusion, and a snare, into which, with a fatal certainty, they would be drawn and destroyed. So cruel at first sight would appear to be the law of nature, as regards the individual; but so merciful when we consider the welfare of the race.

NON-RESTRAINT IN THE TREATMENT OF THE INSANE.

THE tomb of St. Dympna, the patron saint of the remarkable lunatic colony at Gheel, in Belgium, is sought to this day by the faithful, who have worn away for ages the stones surrounding her effigy in their prayers for her propitiatory influence on behalf of their afflicted friends. And on that spot, at least, it may be said that her influence has not been unfelt. But throughout Europe and for many ages, the treatment of the insane was based on the old priestly conception that madness meant possession by the devil. The awful visitations which darken and overthrow the mind of man were regarded as visible manifestations of the Evil One, to be exorcised by charms or averted by the ritual of superstition. Physical as well as spiritual influences were, however, not forgotten; and the priestly leeches, whilst they inculcated an appeal to the Most High in aid of their efforts to evict the archfiend, did not neglect to employ the devil's own weapons in the form of brutal treatment. But it was left for later times to invent so-called scientific contrivances to wrench madness out of suffering humanity, and especially to German subtilty and imagination to devise methods of torture which transcended any amount of simple physical brutality. Instead of eviction by the grace of God, terror and surprise were called into play. Devices of so devilish a nature were sometimes employed, that we are left to doubt whether the physician or the patients were the most insane. One of these was to entice the sufferers to walk across a floor, that suddenly gave way and dropped them into a bath of surprise, in which they were half-drowned and half frightened to death. A still more demoniacal plan of treatment was sometimes employed. Patients were confined by chains in a well, and the water was gradually made to ascend, thus exposing the poor victims to what appeared to them the gradual approach of inevitable death. But such terror was not sufficiently imaginative or romantic, Dr. Conolly tells us, to satisfy some German physicians who " wished for machinery by which a patient just arrived at an asylum, and after being drawn with frightful clangour over a metal bridge across a moat, could be suddenly raised to the top of a tower, and as suddenly lowered into a dark and subterranean cavern; and they owned that if the patient could be made to alight among snakes and serpents it would be still better." In England, as late as the middle of the last century, the national tendency favoured mechanical contrivances less mentally terrifying, but even more physically cruel. A Dr. Darwin invented the circular swing, in which monomaniacal and melancholy patients were bound in the longitudinal position when it was required to induce sleep, and in the erect position when intestinal action was required. This instrument was said to produce such results that the mere mention of its name was enough to induce terror. Dr. Cox, a physician, desired to improve upon this swing by advising that it should be used in the dark in hopeless cases, with the addition of unusual noises and smells. Yet this terrible contrivance was regarded by physicians, of, we presume, ordinary humanity with such approval that it is spoken of by Dr. Hallaran as an invention that no well-regulated asylum should be without. A curious example this of the complacency of even educated men in accepting arrangements, however cruel, with which they are familiar, and a warning to asylum physicians of this age to beware of what Bacon calls the " Idols of the den."

We confess that it is painful and perhaps unnecessary to trace back so far the misery the insane have undergone; and we should not have continued the sad story, were it not advisable to show that the judicious treatment of the insane is a progressive science nobly developed by our fathers and contemporaries, but yet capable of a still wider extension by our sons, labouring in a season when the fair humanities give promise of sweeping away like a flood all the old ideas which in a modified form still surround asylum life.

The evidence given by witnesses before committees of the House of Commons in 1815, relative to the condition of the old York Asylum and of Bethlehem Hospital, shows that within the memory of living men patients were treated more like furious beasts than human beings. In the latter asylum they were shown to the public on certain days of the week, the charge being only twopence, a less sum than it cost to see the lions in the Tower. It was the custom for the blackguards of the town, and even for women, to jeer and mimic the demented inmates in order to excite them to rage. Refractory patients were heavily chained; sometimes those who were not violent were fastened like savage dogs to the wall. Mr. Wakefield, reporting his visit, said :—

"Attended by the steward of the hospital, and likewise by a female keeper, we proceeded to visit the women's galleries. One of the side rooms contained about ten patients, each chained by one arm or leg to the wall, the chain allowing them only to stand up by the bench or form fixed to the wall, or to sit down to it. The nakedness of each patient was covered by a blanket-gown, only the blanket-gown was a blanket formed something like a dressing-gown, with nothing to fasten it in front; this constitutes the whole covering; the feet were even naked."

In another part of the house many women were found locked up in cells, naked and chained, on straw, with only one blanket for a covering; but this being the common treatment at the time, did not seem to strike the public mind so much as the case of William Norris, whose figure may be said to stand out as a martyr and a liberator, for the atrocious treatment of this poor creature not only roused the indignation of the whole British community, but was instanced as a terrible example of our treatment by foreign physicians—very unfairly, by the way, inasmuch as the Retreat at York, instituted and supported by the Quakers, which exercised less restraint than any other asylum in Europe, had been in operation long previous. Bethlehem, however, being the most noted public asylum in the metropolis, naturally attracted more attention than any other. Norris, it appears, was at times violent, no doubt in consequence of the indignities he had to put up with from his drunken keeper. In order to control him, it was suggested by the apothecary that he should be chained, and that the chain should be passed through a hole in the wall of his cell, so that when it was necessary to approach him, he might be hauled up by the chain. Luckily, want of room would not permit of the acceptance by the governors of this wild-beast treatment, and a more economical cage as regards space was contrived for him, which is thus described by the French Asylum physician, Esquirol:—

"A short iron ring was riveted round his neck, from which a short chain passed to a ring made to pass upwards and downwards on an upright massive bar more than six feet high, inserted into the wall. Round his body a strong iron bar about two inches wide was riveted; on each side of the bar was a circular projection, which, being fastened to and enclosing each of his arms, pressed them close to his side."

Thus manacled he lived for nine years. It is noteworthy, as showing the dangerous influence of an asylum atmosphere, that the Committee of Governors of the hospital, in their report upon the evidence given concerning this infernal contrivance, state that "it appears to have been upon the whole *rather a merciful and humane than a rigorous and severe imprisonment!*" And as a proof that it was so, they affirm "that he never complained of its having given him pressure or pain!"

Dr. Munro, the chief physician, who gave his assent to the use of this cruel cage, and under whose care the poor women were chained to the walls in the different wards, stated before the Committee that "irons were only fit for paupers; that they were never used for his own private patients." . . . Being asked why a gentleman would not like irons, his reply was indicative of a social contempt of the lower classes which seems strange enough at the present day, especially after the loving tenderness of Conolly for the poor and neglected. " In the first place," replied he, " I am not at all accustomed to gentlemen in irons ; I never saw anything of the kind ; it is a thing so totally abhorrent to my feelings, that I never considered it necessary to put a gentleman in irons." But the highest rank did not exempt the unhappy victims of mental disease from treatment at which humanity recoils. Mr. Massey, in his " History of King George III.," has drawn from the Harcourt Papers an affecting picture of the atrocious treatment to which the King was subjected in 1788, when Dr. Warren regarded him as a confirmed lunatic. The King's disorder manifested itself principally in unceasing talk (he talked once for nineteen hours without intermission), but no disposition to violence was exhibited. Yet he was subjected constantly to the severe restraint of the strait-waistcoat; he was secluded from the Queen and his family, and denied the use of a knife and fork. He was abandoned to the care of low mercenaries, one of whom— a German page named Ernst—actually struck him. The King, after his recovery, retained a lively recollection of these outrages. No sooner was Dr. Willis called in than all this changed. That estimable person immediately soothed his patient, released him from coercive restraint, presented him with a razor to shave himself, and when the King demanded a knife and fork he courteously assented, saying, that he hoped to be allowed the honour of dining with his Majesty. The Queen and Princesses were again brought into his presence. These measures were viewed with the greatest jealousy and alarm by the Court physicians, but the consequence was that the King in a few weeks entirely recovered.

That was one of the first and most striking instances of a victory gained by non-restraint over madness.

The effect of the parliamentary inquiry of 1815 was exceedingly great. It struck the first great blow at the bad experience which is the bane of lunatic establishments. The periodical vomitings and purgings which at stated times were indiscriminately administered to the patients, regardless of necessity, but because Dr. Munro had inherited the practice from his father, were given up; poor Norris was extracted from his iron cage, and after having been so long confined in it, to the prone or erect position, thankful for small mercies, expressed his thanks that he was "allowed to sit down on the edge of his bed." The poor women that hung from their fetters and chains on the wall, like vermin chained to a barn door, were liberated, dressed like human creatures, and became at once calmer;

and Dr. Haslam, the apothecary, who was the medical despot of the hospital, notwithstanding his proud boast to the Committee, "I am so much regulated by my own experience that I have not been disposed to listen to those who have had less experience than myself" (a remark we sometimes still hear, by the way, from asylum superintendents), found that the fresh breath of a humane public opinion had blown to the winds his cruel conceit, and so changed the den that he had "hung with curses dark," that visitors, horrified but a year before by the sights and sounds in the asylum, now scarcely recognized it, so changed and quiet were the wards.

In all public asylums, and wherever any public supervision penetrated, chains were abolished, and to this extent the poor insane pauper was put upon a par with the gentleman; but handcuffs and strait-waistcoats were still considered implements that "no well-regulated asylum should be without." The time was at hand, however, in which the force of public opinion, even in respect to these minor forms of personal restraint, was about to influence old ideas. In 1803 an article on Pinel's "Aliénation Mentale," written by Dr. Henry Reeve, who was afterwards physician to the Norfolk and Norwich Bethlehem Hospital, where he introduced a milder form of treatment, appeared; and a spirited review by Sydney Smith of Tuke's work on Non-Restraint, published in 1814, contributed to enlarge the notions of resident physicians of asylums with respect to this great principle, which before long was to receive a larger practical development at their hands. Still it was accident again that gave the next impulse to the movement, and this took place in the Lincoln Asylum. Conolly, in his "Treatment of the Insane," tells us that "a patient in that asylum had died in the year 1829, in consequence of being strapped to a bed in a strait-waistcoat during the night; and this accident led to the establishment of an important rule, that whenever restraints were used in the night, an attendant should continue in the room; a rule which had the desired effect of much diminishing the supposed frequency of such restraints being necessary." It was soon found that a principle that answered so well at night was also applicable by day, and the consequence was, that by degrees the necessity for restraint became less frequent, so much so, that for some successive days the asylum records were without any entry of their use. This was in the year 1834, at which time Mr. Hadwin was the house-surgeon of the asylum. In 1835, Mr. Gardner Hill succeeded him. Imbued with the spirit of his predecessor, he still further ignored the use of mechanical restraint, and in 1837 he boldly declared that it might be altogether abolished.

As the name of Dr. Charlesworth, the visiting physician to the Lincoln Asylum, has been associated with that of Dr. Gardner Hill as an equal labourer in carrying out the new idea—nay, has been placed by some as the real discoverer—we think it but fair that the evidence furnished by Dr. Gardner Hill in his volume, " Lunacy, Past and Present," should be adduced, and we hold it to be conclusive. Whilst it must be admitted that Dr. Charlesworth readily received the ideas of the house-surgeon of the Lincoln Asylum, and warmly seconded him in his bold attempt to throw away all implements of restraint, it cannot be further maintained that he had any right to the name of inventor of the system. Dr. Conolly, indeed, refers to him as sharing with Dr. Gardner Hill that credit, but this must be ascribed to a too partial friendship. Dr. Gardner Hill is certainly not persuasive in his style, and for this reason has raised up many enemies to his assertions; but truth compels us to say that the following evidence of his claims to the great honour of being the first to do away with mechanical methods of restraint is indisputable. The report of the Lincoln Asylum for 1836 refers thus early in the history of the great experiment to the success of Dr. Gardner Hill's fruitful idea :— " Three successive months (except one day) have now elapsed without the occurrence of a single instance of restraint in this establishment; and out of thirty-six weeks that the house-surgeon has held his present situation, twenty-five whole weeks, excepting two days, have been passed without any recourse to such means, and even without an instance of confinement to a separate room." Again, in the report of 1838, which is signed by the

Chairman of the Visiting Committee, E. P. Charlesworth, the merit of the new idea is unequivocally ascribed to the house-surgeon — no mention being made of Dr. Charlesworth's name. "There is now," says this report, "an increased confidence that the anticipations of the last year may be fulfilled, and that an example may be offered of a public asylum, in which undivided personal attention towards the patients shall be altogether substituted for the use of instruments of restraint." "*The bold conception* of pushing the mitigation of restraint, of actually and formally abolishing the practice mentioned in the last report, due to Mr. Hill, the house-surgeon, seems to be justified by the following abstract of a statistical table, showing the rapid advancement of the abatement of restraint in this asylum under an improved construction of the building, night-watching, and attentive supervision." The table thus mentioned shows that the number of hours passed by patients under restraint diminished from 20,423 in 1829 to a significant 0 in the year 1838. Although Dr. Charlesworth heartily seconded his endeavours, and for so doing deserves great praise, yet it was not to be supposed that so mighty a reform could be effected without the opposition of the usual number of obstructives to all original ideas. Dr. Hill says :—

"Within the walls I had the whole staff of attendants against me. I prevailed over the attendants by going amongst them and personally superintending the refractory patients. I spent several hours daily in the disorderly patients' wards for weeks in succession—in fact I watched the attendants and the patients until I felt satisfied that restraint was a pretext for idleness, and nothing more. When restraint was abolished, then ceased the reign of 'guttling, guzzling, and getting drunk by the attendants,' as had been the case under former management. Outside the asylum I had the whole medical world against me. The superintendents of several of our largest asylums opened a regular battery against me. I was assailed right and left. The 'Hillite system,' as they pleased to term it, was denounced as 'Utopian.' By one it was called 'an absurd dogma,' by another 'a gross and palpable absurdity;' some fulminated against it as 'the wild scheme of a philanthropic visionary, unscientific, and impossible;' by others as the ravings of a theoretic philosopher, involving the unnecessary exposure of the lives of the attendants—in fact, *a practical breaking of the Sixth Commandment*. Others, more moderate in their views, denounced it as speculative, peculative, &c. &c. Dr. Clutterbuck rhetorically condemned it 'empirical, and highly dangerous to the patient and to those around him.' Dr. James Johnston said 'it indicated insanity on the part of its supporters; it was a mania which, like others, would have its day;' and Monsieur Moreau de Tours said that 'the idea was entirely Britannic; that it was *an impossibility in most cases*, an illusion always, and the expression itself a lie.'"

It seems very hard indeed if, after all these rough words, the medical man who called them forth should be deprived of the merit of having given occasion for them!

Thus, in the words of Dr. Conolly, the non-restraint system became established at Lincoln. It is to the infinite credit of the noble nature of the great reformer, that he never failed to admit, especially in public, that the initiative of the new system was not due to himself. To Dr. Gardner Hill this great merit was due; to his lectures, indeed, on lunatic asylums, delivered at the Mechanics' Institution at Lincoln in 1838, Dr. Conolly owed the happy inspiration which led him to embrace the new doctrine. In order to convince himself of its truth, before he assumed the post of resident physician at Hanwell Asylum, he visited the Lincoln Asylum and witnessed its practical application.

It must strike many minds that the world has dealt unfairly in practically ignoring, as it has done, the claims of Dr. Gardner Hill. In all great discoveries it is generally the one who has translated ideas into acts that has reaped the final reward. The great Pinel, Dr. Tuke of the York Asylum, Dr. Hadwin of the Lincoln Asylum, all contributed their stone to the new idea, but it is to

Hill that the undoubted claim of courageously clearing an asylum of all mechanical implements of restraint is incontestably due; and for this service the crown that is due to him should no longer be withheld. And this may be done without taking one inch from the stature of Conolly, who so modestly repudiated any claim to the idea during his life.

But to Conolly belongs a still higher crown, not merely for his courage in carrying out a beneficent conception on a large scale and on a conspicuous theatre, but for his genius in expanding it. To him, hobbles and chains, handcuffs and muffs, were but material impediments that merely confined the limbs; to get rid of these he spent the best years of his life; but beyond these mechanical fetters he saw there were a hundred fetters to the spirit, which human sympathy, courage, and time only could remove.

Perfect as was the experiment carried out at Lincoln Asylum, the remoteness of that institution from the great centre of life, and the want of authority in its author, would no doubt have prevented its acceptance for years by the physicians of the great county asylums so long wedded to old habits. It was for some time treated as the freak of an enthusiastic mind, that would speedily go the way of all such new-fangled notions; and no doubt it would, had not an irresistible impulse been given to it by the installation of Dr. Conolly at Hanwell, where, with a noble ardour, he at once set to work to carry out in the then largest asylum in the kingdom the lesson he had learned at Lincoln.

Dr. Henry Maudsley, in his sketch of the life of Conolly, in the *Journal of Mental Science*, dwells upon the feminine type of his mind:—" A character most graceful and beautiful in a woman, is no gift of fortune to a man having to meet the adverse circumstances and the pressing occasions of a tumultuous life." Now and then humanity has to thank the Creator for the seeming imperfections of His creatures. No doubt this great reformer's mind was not of the self-contained perfect type that would have satisfied Mr. Carlyle; it was, on the other hand, utterly lopsided: more womanly than the mind of a woman, it seemed to begin and end with love and sympathy: and what a world of sympathy it requires to deal with the demented, fatuous, and idiotic, those only know who have been brought into constant contact with them. Together with Pinel, the great French psychologist, he possessed the rare gift of moral courage, or rather, shall we say, he possessed a firm belief in the power of gentle and humane feeling to conquer the most outrageous passions. Notwithstanding the tremendous responsibility

both these noble men took upon themselves
when they unloosed the bonds of their prisoners, they never hesitated, or doubted of
the result of the step they were about to take.
They were alike discouraged. " Experience,"
that dreadful impediment to all progressive
science, shook its head doubtfully, and antici-
pated their discomfiture. Couthon, in 1792,
after interrogating, at the request of Pinel,
the inmates of the Bicêtre, whom that philanthropist proposed to reclaim, recoiled
with horror from the proposal. "You may
do as you please with them," said he;
"but I fear you will become their victim."
In the same manner Conolly's attempts were
met with incredulous pity. His "want of
experience" in lunatic asylums was ever
quoted against him ; and after the success of
the system of non-restraint was proved, the
superintendents of other asylums were still
unbelievers. In a letter to Mr. Hunt, of

Stratford, recording his success, he says:—
"Our asylum is now almost daily visited by the officers of other institutions, who are curious to know what method of restraint we *do resort to*, and they can scarcely believe that we rely wholly on constant superintendence, constant kindness, and firmness when required."

It is very curious to note the difference with which Pinel and Conolly reviewed the first results of their brave work — the dramatic detail of the Frenchman with the calm narrative style in which the physician of Hanwell describes the relief from bonds of a whole asylum full of lunatics:—

"The first experiment of Pinel was tried upon an English captain, whose history no one knew, as he had been in chains for forty years. He was thought to be the most furious among them; his keepers approached him with caution, as he had, in a fit of fury, killed one of them on the spot with a blow from his manacles. He was chained more rigorously than any of the others. Pinel entered his cell unattended, and calmly said to him, 'Captain, I will order your chains to be taken off, and give you liberty to walk in the court, if you will pormise me to behave well and injure no one.' 'Yes, I promise you,' said the maniac; 'but you are laughing at me; you are all too much afraid of me.' 'I have six persons,' answered Pinel, 'ready to enforce my commands if necessary. Believe me then, on my word, I will give you liberty if you will put on this strait-waistcoat.' He submitted to this willingly, without a word; his chains were removed and the keepers retired, leaving the door of the cell open. He raised himself many times from the seat, but fell again on it, for he had been in a sitting position so long that he had lost the use of his legs; in a quarter of an hour he succeeded in maintaining his balance, and with tottering steps he came to the door of his dark cell. His first look was at the sky, and he exclaimed enthusiastically, 'How beautiful!' During the rest of the day he was continually in motion, walking up and down the staircase and uttering exclamations of delight. In the evening he returned of his own accord to his cell, where a better bed than he had been accustomed to had been provided for him, and he slept tranquilly. During the two succeeding years which he spent in the Bicêtre, he had no return of his paroxysms, but even rendered himself useful by exercising a kind of authority over the insane patients, whom he ruled in his own fashion. In the course of a few days Pinel released fifty-three maniacs from their chains; among them were men of all conditions and countries,—workmen, merchants, soldiers, lawyers, &c. The result was beyond his hopes; tranquillity and harmony succeeded to tumult and disorder, and the whole discipline was marked with a regularity and kindness which had the most favourable effect on the insane themselves, rendering even the most furious more tractable."

But this humane conduct nearly cost him his life. The Paris mob did not believe in his humanity, and attributing it to some base motive, seized him one day in the streets, and would have hung him but for the interference of an old soldier of the guard, whom he had liberated from his chains.

The English physician, although quite as enthusiastic as Pinel, is still ruled by national calmness of thought, and his account of the first four months of non-restraint as experienced at Hanwell, is given in a letter to his friend, Mr. Hunt, of Stratford, in a manner so quiet and undemonstrative, that the greatness of the experiment seems lost in the simplicity of the record. Not only had he to deal with a much larger number of lunatics than Pinel—there were eight hundred at Hanwell when he made his first venture— but when he loosened their bonds he had no strait-waistcoats and other articles of restraint, like the physician of the Bicêtre, to fall back upon. What he gave was absolute freedom, as far as the use of the limbs was concerned; and had he resorted to even the slightest means of mechanical control, the enemies of the new movement, who were jealously watching him, would have declared that he had failed. Under such circumstances, the humble spirit in which he announces his triumph is very remarkable :—

"I know you will feel glad," he says, writing to his friend in January, 1840, " that we have now ruled this great house for four months without a single instance of restraint by any of the old and objectionable methods. The use of strait-waistcoats is abolished, hand-straps and leg-locks never resorted to, and the restraint-chairs have been cut up to make a floor for the carpenter's shop. All this of course occasioned some trouble and some anxiety, but the success of the plan and its visible good effect abundantly repay me. I think I feel more deeply interested in my work every day. I meet with the most constant and kind support of the magistrates; indeed, my only fear is that they should say too much of what is done here, and thus provoke envy and censure."

Looking at the matter as we now do, so long after the practical process of the abolition of all means of personal restraint has been established, we cannot fairly estimate the anxiety of mind that must have oppressed Conolly, when having thrown away the fetters he stood face to face with suicidal patients whose great aim in life is to get rid of it. The enduring cunning of this class of patients in compassing their ends, their adroitness, their impulsive vigour, but too well known to him, must have been before him night and day—a single life lost at this moment of trial, and the whole superstructure would have crumbled to the dust. It unfortunately happened that during the second year of trial nine such cases were brought to Hanwell; many of them came in a raving condition, bound hand and foot; they were taken to the wards and then set free, whilst those who brought them fled in terror. Well might the resident physician, in the presence of such crucial tests of the faith that was in him, tremble for its success. Instead of rigid bonds to confine the patient's limbs, he had nothing to resort to but unceasing watchfulness and sympathy. These were to all the world but himself weak and impotent substitutes; but the event proved that he looked with larger eyes than his contemporaries, and his courage was responded to with the most complete success. The abolition of all means of personal restraint was soon found to have more than a temporary influence upon the patients. It modified the very types of insanity. Instead of calming the patients, bonds only exasperated them, and their features, from their constant employment, settled into rigid expressions of rage and fury, that we are only familiar with in the prints of madhouse scenes in the old times—to wit, Hogarth's grim sketches, which seem almost to caricature human nature, even when exhibiting the most diabolical expressions. Conolly in his fifth report notices this extraordinary change:—

"Fresh illustrations have been daily afforded of the advantages of those general principles of treatment, which have been expressed in former reports, and of which the effects are to remove as far as possible all causes of irritation and excitement from the irritable ; to soothe, encourage, and comfort the depressed ; to repress the violent by methods which leave no ill effects on the temper, and leave no painful recollections on their memory; and in all cases to seize every opportunity of promoting a restoration of the healthy exercise of the understanding and of the affections. Insanity thus treated undergoes great, if not unexpected, modifications ; and the wards of lunatic asylums no longer illustrate the harrowing descriptions of their former state. Maniacs not exasperated by severity, and melancholy not deepened by the want of all ordinary consolation, lose the exaggerated character in which they were formerly beheld."

The history of the four months from the 1st of June to the 31st of October, 1839, the date of the first report, presented to the Quarter Sessions, by the resident physician of Hanwell Asylum, repeats itself in all the subsequent reports from his pen. Implements of coercion were abolished once and for all; and if the history of non-restraint was limited to a mere record of the disuse of these mechanical implements, the record would be very slight indeed; but, as we have before said, Conolly took no such limited view of the great theme he was handling. In his mind non-restraint was synonymous with an entire absence of any circumstance or thing that unnecessarily irritated or thwarted the patient—a position asylum physicians, as a rule, have not yet fully comprehended. There are methods of coercion which wound the spirit still more than manacles hurt the body. Fully aware of the tyranny that may be inflicted without the use of iron or thong, in every page of his works he enforces the necessity for human sympathy and kindness. That the philanthropic labours of Dr. Conolly were not overlooked by his contemporaries we have proof in the following extract from the first number of the *Psychological Journal of Medicine*, written by the editor (Dr. Forbes Winslow) in the year 1848. In reviewing Dr. Conolly's work on the "Construction and Government of Asylums," Dr. Winslow thus bears honourable testimony to that physician's benevolent exertions on the behalf of the insane then under his care in the Hanwell Asylum :—

"Let the hundreds who annually visit this noble institution, and wend their way through its wards, inspect its arrangements, and perambulate through its grounds, give evidence of the admirable skill with which everything is conducted. Dr. Conolly's spirit appears to pervade every department of the asylum; he is like a father among his children, speaking a word of comfort to one, cheering another, and exercising a kindly and humane influence over all; making the very atmosphere in which the patients live redolent of the best sympathies of our nature. He feels, as all ought to feel who undertake the important, the anxious, and responsible management of the insane, *that the affliction of disease does not necessarily block up the avenues to the human heart;* that even in the worst, the most distressing forms of mental malady, there often exist some of the better principles of our spiritual being in all their original purity, upon which the physician and the moralist may act with advantage."

In this liberal and just view of the treatment of the insane, we fear he has left but few disciples behind, few who see the whole scope of his system, or at least have courage to carry it out. Had he lived, he would not have thought that the county asylum was the latest expression of his idea, or have contented himself with that form of brick-and-mortar humanity which county magistrates so affect. Indeed, we have his own words in condemnation of asylum extension, at a time when it had not reached its present monstrous development. Many years ago, in a letter addressed to Sir James Clark, he says :—

"In the monstrous asylums of Hanwell and Colney Hatch, sanitary principles have been forgotten and efficient superintendence rendered impossible. The magistrates go on adding wing to wing and storey to storey, contrary to the opinion of the profession and to common sense, rendering the institution most unfavourable to the treatment of patients, and their management most harassing and unsatisfactory to the medical superintendent."

And this process of enlargement is going on with redoubled vigour all over the kingdom. Nearly every county asylum is demanding and obtaining enlargement, and applicants are overtaking even these enlargements. It is capable of proof that lunacy is not increasing in a greater ratio than the population, but still they flow into these asylums quicker than the old inmates die out. The very imposing appearance of these establishments acts as an advertisement to draw patients towards them. If we make a convenient lumber-room, we all know how speedily it becomes filled up with lumber. The county asylum is the mental lumber-room of the surrounding district; friends are only too willing, in their poverty, to place away the human encumbrance of the family in a palatial building, at the county expense. But though the natural appearance of these institutions is so attractive, the pleasure-grounds look so well kept, the walks so trim, everything that is merely material is in such good order, we fear that we must demur to the extravagant opinions that have been uttered with respect to their qualifications as places of mental cure. Insanity does not wholly alter a man's nature: as a rule, his old instincts, habits, and feelings remain exaggerated or twisted in some cases no doubt, but still they form an integral part of his nature, and cannot be rudely violated or oppressed without creating natural offence.

Let us enter one of these fair asylums however, which, according to Professor Paget of the Cambridge University, "is the most blessed manifestation of true civilization that the world can present." Let us, as we have said, pass along these interminable wards and examine this paradise which rouses the Professor to such an enthusiastic approval—enter not with heart hardened by long endurance and deadened by that dreadful experience, which kills all attempts at reform; but with a fresh mind which does not refuse the lunatic in his harmless condition at least some of the ordinary feelings and emotions of our common humanity. The first thing that strikes us is the monastic and cloistral system which obtains. It would appear as though it were an offence in asylum life for men and women to meet together. We all know the amenities that prevail in convent life, and of the manner in which nuns love one another; how then can we wonder that the female patients we pass in the long galleries are eaten up by utter vacuity and dreariness; or that the men, only a stone's throw off, herd hopelessly together, starved of some of the best feelings of ordinary life, such as arise from social intercourse with the other sex? It strikes one with astonishment to see the airing courts thus sorted as if especially to make the wanderers miserable; to see that even meals cannot be taken in common. We ask in vain why this unnatural division is established — a division which, while it violates nature, deprives the physician of one of his best means of cure. Some years ago it was the custom of Colney Hatch for the females and males to dine in one room, but at different tables—an expedient which at the time called forth the praise of the Visiting Commissioners; but even this mild, not to say aggravating, approach to a more natural state of things—at a distance, has of late been discontinued. There is no objection urged against a natural mingling of the sexes under proper precautions, and the only practical objection urged against it that we have ever heard, is that the organization of asylums does not permit of these mixed gatherings. The decorous and regulated intercourse of the sexes is in itself an invaluable lesson in self-restraint.

Towards the end of Dr. Conolly's life, he was oppressed with many fears lest the advance that had been made should, through the selfishness and neglect of mankind, lose its impulse, and indeed be permitted to go back. The present age is certainly not less philanthropic than the one in which he carried out this great reform, but there are certain elements at work in asylum life that justify some of his apprehensions. The first and foremost of these is the gradual growth of the county asylums. Some of these have become so large that anything like individual treatment of the patients is quite out of the question. They have ceased to be houses for the cure of mental disease, and have subsided substantially into mere houses of detention. And not only have they outgrown their curative capabilities, but they have also degenerated from that high standard as houses of mercy and pity, to which Conolly would have them raised. No one saw more clearly than that philanthropist the fact that the abolition of all means of mechanical restraint put the asylum physician at the mercy of his attendants. In place of the strait-waistcoat, which, with all its faults, acted without passion, he had to rely upon human force liable to human weakness. To keep this in check the most careful supervision is absolutely necessary—a supervision on the part of the medical officers, that is ludicrously inadequate on account of their limited numbers: the result is that, as a rule, the patients are at the mercy of the attendants. What that mercy is, let the inquests that have been held in asylums on patients who have died through brutal ill-treatment at their hands make the sad answer. We do not wish to be hard upon these "instruments of the physician's will," as Conolly terms them; they are neither better nor worse than others in the same class of life: those only who know how trying are their duties can fairly make sufficient excuses for them; but as a fact, the school they go to is not calculated to teach humanity to uneducated minds, and we more than fear they do not forget their instruction. What we say is no mere surmise. The difficulty of obtaining trustworthy attendants is one of the trials of the medical superintendent. Yet, without their intelligent aid, he works in the dark.

"The physician," says Conolly, "who justly understands the non-restraint system, well knows that the attendants are the most essential instruments, that all his plans and all his care, all his personal labour, must be counteracted if he has not attendants who will observe his rules when he is not in the wards as conscientiously as when he is present."

Again, he says, significantly enough :—

"Attendants are generally persons of small education, and easily inflated by authority; they love to command rather than to persuade, and are too prone to consider their patients as poor lost creatures, whom they may drive about like sheep."

We fear the attendants of the present day are not one whit improved. There are certain asylums that have such a bad name for those trained in them, that they stand no chance of obtaining service with the medical superintendents of other establishments. Indeed, such are the tricks they learn, that many asylum physicians prefer obtaining assistants who have never seen asylum life. As the strength of any establishment must be measured by its weakest part, we fear that county asylums in this particular come off but very poorly. As we have said before, it is the attendant who is the real master of the patient: hour by hour he is at his mercy. The many small cruelties he perpetrates, sometimes from temper; the many neglects he is guilty of, often in consequence of fatigue, are seldom known and are but rarely recorded. It is only when some dreadful cruelty happens that the world is made cognisant through an inquest, that restraint has not altogether vanished with the destruction of bonds. When we hear, as we have too often of late, that a poor demented creature has had his ribs crushed in by the knees of his attendant whilst kneeling upon him, or trampling on his chest in that position, possibly the public might be induced to think twice over the verdict, that "the county asylum is the most blessed manifestation of true civilization that the world can present." Not long since no less than three convictions were obtained in different parts of the country against the keepers of lunatics for acts of brutality and violence. No wonder Lord Shaftesbury expresses a hope that these verdicts may have a salutary effect in future.

At all events poor Reynolds, who died whilst experiencing one of the "manifestations of civilization," would have been able to put in his protest against this doctrine if they had only given him a little more time to live. For these evils the county magistrates are wholly answerable. The Visiting Commissioners have over and over again protested against the enlargement of asylums, clearly seeing as they do, that the whole spirit of non-restraint is thereby contraverted, but unhappily the Commissioners have no power to avert the evil. The supervising power established by the Government to correct the tendency to slip back into restraint, is set at nought by the jealousy of the county magistrates, who hold the purse-strings. With them the county asylum is mainly an institution to maintain and keep lunatics on the club system, and their cure, the only proper object of an asylum in the eyes of the physician and the legislature, is made a secondary object.

" If," says Dr. Conolly, "the public would really estimate the consequences of the present inadequate number of medical officers in relation to their duties, which at least ought to be performed in asylums, an augmentation would be insisted upon. With the various interruptions to which they are liable, it is quite evident that the medical officers cannot sufficiently superintend a thousand patients; that they cannot even sufficiently visit the wards often without exhaustion, and consequently cannot exercise due supervision over the attendants; that on numerous occasions important duties must be omitted, and important circumstances overlooked, and that many special moral appliances must be neglected with serious consequences, not the less real because they are unrecorded. Without a very efficient superintendence, chiefly to be exercised by the medical officer, or rather by the chief medical officer, the mere absence of mechanical restraint may constitute no sufficient security against the neglect, nor even the actual ill-treatment, of insane persons in a large asylum. The medical officers who consider such watchful superintendence not properly comprised in their duties, have formed a very inadequate conception of them."

The absurd rules which are forced by the magistrates on the medical superintendents take up much of the little time they have for their overwhelming daily labours. We were informed by one of these gentlemen, that by the rules of his asylum, he was obliged to make an entry of his visit every time he entered a ward; and this piece of needless clerkship alone occupied forty minutes every day. Whilst we dwell with pride upon the fact that mechanical restraint is practically abolished in this country, let us not forget that foreigners sometimes regard with astonishment the miserably inadequate staff with which we are contented to work our asylums. Colney Hatch, with its 2,000 patients, has only four medical officers,—is it to be wondered at that foreign physicians refuse to believe in our boasted moral treatment when they find our medical supervision so miserable a sham?

The patients are treated on an organized system, very well suited to a workhouse, but totally unfitted to an asylum for mental cure Individuality is entirely overlooked; indeed, the whole asylum life is the opposite of the ordinary mode of living of the working classes. When the visitor strolls along the galleries filled with listless patients, the utter absence of any object to afford amusement or occupation strikes him most painfully. Care is taken to shut out the ever-varying scenes and passages of life, so full of variety and so fraught with interest. Every natural emotion and healthy motive that freshens the intercourse between human beings in the outside world is excluded from them; and what is substituted? It is remarked with infinite approval now and then by the Commissioners that the walls have been enlivened with some cheap paper, and a few prints have been hung in the galleries, that a fernery has been established—matters all very well in their way, but utterly inadequate to take the place of the moving sights and scenes of the outside world. Can we wonder that the chronic and convalescent patients grow weary of their prison, that the very sight of the asylum is hateful to them, that the greatest treat you can give them is a walk out of sight of its walls?

The great want admitted in every asylum is occupation. In the county asylums the labourer goes with a sense of relief to work at the farm, and the artisan takes his place in the workshops—those true places of cure when moderately used. But even these invaluable aids to medicine may, we think, be greatly improved. At present by many patients the work is looked upon as mere diversion, it lacks the stimulus that urges on a man in the world. As it is admitted that the object in setting the patient to work is not that he may repay by his labour the cost of his treatment, but that he may be induced to cast aside his hallucinations and fancies, and return once more to healthy feelings and thoughts—why is the healthy stimulus of pay withheld? How many a man would be gradually drawn from his insanity if he knew his labour was to have its reward, and that he would leave the asylum with help for those his illness had thrown into sore poverty and distress!

The time has at length arrived when it is obvious that if our asylums are to resume the true position from which they should never have been allowed to depart—that of hospitals for the treatment of the insane—a thorough revolution must be made in their management; and in order to bring about new measures, we must pray for the advent of entirely new men. There are epochs in all institutions at which a paralysis seems to seize upon those conducting them. With regard to our present superintendents as a body, with a few noble exceptions, we unhesitatingly assert the spirit of Conolly is dead. A miserable spirit of routine, without resources, spring, or energy, is sapping and destroying asylum life. The gross fallacy of supposing that no man without experience in pauper lunatic asylums is capable of taking charge of such establishments, is the cause of an infinity of mischief. Our own belief is, that wholly fresh blood is imperatively demanded. Who have been the great reformers—the leaders in the onward, ever onward course of non-restraint? Not physicians trained in all the bad traditions of asylums, but general physicians, who have come to the task with fresh minds and habits untainted by an unhappy experience. Pinel, before he took charge of the Bicêtre, was a general physician. Conolly, happily, was innocent of the ways of asylums before he became superintendent of Hanwell; and the far-famed Retreat at York received its inspiration from an intelligent Quaker layman, William Tuke, of York. It is the same with all other professions and arts; improvements come, as a rule, from without; from a class of thinkers, who have not to unlearn habits of mind instilled into them by a kind of Chinese practice and a reverence for old authority.

No doubt in the eyes of the public these establishments are the necessary places of detention of troops of violent madmen, too dangerous to be allowed outside the walls. It is difficult to get rid of old notions on the subject of lunatics. The popular idea is that they must all be raving and desperate, and the visitor to an asylum enters the wards with the expectation of meeting violent maniacs, whom it would be dangerous to approach. He has not taken many steps, however, before this illusion begins to vanish; he may even ask, " Where are the mad people ? " as he sees nothing but groups of patients seated round the fire or lolling about in a dreary sort of way, perfectly quiet, and only curious about the curiosity of the stranger. This is the class of people that form at least 90 per cent. of the inhabitants of our asylums, chronic and incurable cases that no treatment will ever improve, upon whom the elaborate and expensive classification and organization is entirely thrown away, and to whom the palatial character of the building in which they are immured, not only affords no delight, but is perfectly detestable. It is this class of patients, beyond human help, that now choke up the public asylums throughout the land, converting them from houses of cure into mere prisons. It will doubtless surprise the reader to be informed that out of the total number of 24,748 pauper-patients in county and borough asylums, and in registered hospitals, in the year 1867, no less than 22,257 were past all medical cure, whilst the curable amounted only to

2,491, or a little more than 10 per cent. When we consider the pressure put upon the ratepayers for the erection of large asylums throughout the land, this result is so disastrous that it may be said our whole scheme for the cure of lunatics has utterly broken down. And the mischief is growing from day to day, for the chronic cases are eating up the miserable percentage of beds still open for newly-arrived acute cases. As the asylums are extending in size, the very atmosphere within the walls may be said to be saturated with lunacy. They are becoming centres for the condensation and aggravation of the malady, rather than places of cure; just as the crowding a fever hospital makes the type of disease more malignant. We are convinced that this is an evil that has been too much overlooked. The insane not only require more physical support than the sane to keep them from going back, but also more healthy mental stimulus; they cannot lean upon themselves without deteriorating. Hence the true principle of cure for the curable, and of support for the incurable, *is an association with healthy minds.*

It must not be supposed that the insane are altogether wanting in observation, or that they are uninfluenced by example. To drive weak and perverted minds into a crowd, and there keep them as a class apart, is clearly against the teachings of common sense, and is opposed to scientific observation; and to keep them there unnecessarily is a crime. The most painful impression left upon us after visiting a county asylum is the doleful wail from the patients as they pray for liberty from the medical attendant— all the more painful as we are aware that large numbers are needlessly detained. Of the ninety per cent. of chronic cases, at least thirty, by the admission of the medical superintendents, and probably nearly forty to less official views, are both harmless and quiet, capable of giving some little help in the world, and with a capacity for enjoyment. To deny them their liberty under these circumstances is both cruel and illegal, inasmuch as the certificate of lunacy which is the authority for a patient's detention states that he must be "a *proper* person to be detained and taken charge of," which certainly cannot be said of these poor harmless and incurable creatures.

Thus it will be seen that more than a third of the beds in existing asylums are improperly filled, and may be cleared, not only with advantage to those needlessly detained, but also to the ratepayers, inasmuch as the room they take up would afford accommodation for these next twenty years for the acute and curable cases which cannot now find admission.

The advisability of opening the asylum gates to this crowd of incurable and quiet cases being granted, as it is, by the Commissioners and the medical superintendents of asylums, the next question is, how to provide for them. The Commissioners, with a pardonable official conservatism, have a pet plan of their own : they are perfectly willing, and indeed desirous of clearing the asylums of every patient that can with safety be permitted more freedom, but they cannot make up their minds to let them go beyond sight of the establishment. Thus in their twenty-first report they refer with approval to the associated cottage system which has been adopted in some of our asylums :—

"In the enlargement of existing county asylums, as well as in erection of new ones, it has been our practice to advocate, as far as possible, the construction for the more quiet and trustworthy patients, especially those employed in the farm, or in the laundry and workshop, of inexpensive associated accommodation, homely in character, and simple in architecture."

"The detached blocks erected in Kent, Devon, Chester, Prestwich, Nottingham, Glamorgan, and Wakefield asylums, and the associated accommodation provided in many others in connection with the laundry and the workshops, have proved most successful, and all our experience points to the advantage which not only the quiet working patients derive from this description of accommodation, but even some of the less orderly and tractable."

The advantage of these associated buildings for patients convalescent but still under treatment cannot be denied; they are valuable stepping-stones to the outside world to which these convalescents are tending, but as regards the crowd of harmless incurable patients, the outlet they could possibly supply would be totally inadequate to the demand. Moreover, they are nothing more than the extensions of the asylum, broken fragments scattered around it, and totally wanting in the freedom that is alone valuable for the mass of chronic and incurable cases. It is impossible to refer to this recommendation of the associated cottage system without recollecting that they have been inspired by a far simpler system that has lasted, with approval, for ages, and one which is as much superior to this weak imitation as a fine picture is to a feeble copy.

We referred at the commencement of this article to the one exception to the cruel treatment of the insane that obtained throughout Europe as late as the first quarter of the present century. Singularly enough, the exception was in the land of municipal liberty—Belgium. The following account, gathered from the *Psychological Journal of Medicine*, is the substance of a report by Dr. John Webster of London, to whose discrimination, fairness, and perfect truthfulness psychological medicine, in England at least, is indebted for this rediscovery of an institution which has had an immense influence in furthering the non-restraint principle in its widest and best spirit.

"What is far more interesting to those accustomed to the bolts and bars, the locks, wards, and high walls of crowded European asylums, is the almost entire liberty accorded to the lunatics resident in the town of Gheel and its neighbouring hamlets, to the number of 1,100, or a little more than a tenth of the whole district. The only building in the nature of an asylum is a structure fitted for sixty patients in the town of Gheel itself, lately erected. Here the patients when they arrive are detained a short time on trial, before they are dispersed among the cottages under the care of the *nourriciers*, or attendants, or caretakers, under whom they subsequently remain. The little army of pauper and other patients, gathered from the whole superficies of Belgium, instead of being stowed away in gigantic asylums, such as Colney Hatch, in which all ideas of life are merged in the iron routine of an enormous workhouse, are distributed over six hundred different dwellings, the major portion of which are small cottages or small farmhouses, in which the more violent or poorer patients are dispersed; and the remainder are situated in the town of Gheel, and are appropriated to quieter lunatics and those who are able to pay more liberally for their treatment. In these habitations the sufferers are placed under the care of the host and hostess, more than three patients never being domiciled under one roof, and generally not more than one. The lunatic shares in the usual life of the family, his occupations and employments are theirs, his little cares and occupations are the same as theirs. He goes forth to the fields to labour as in ordinary life ; no

stone walls perpetually imprison him, as in our asylums.
If it is not thought fit for him to labour at the plough
or the spade, he remains at home and takes care of the
children, prunes the trees in the garden, and attends to
the pottage on the fire ; or, if a female, busies herself in
the ordinary domestic duties of the house. The lunatics, as may be supposed, are not left to the discretionary
mercies of the host and hostess. A strict system of
supervision prevails, somewhat analogous to that of the
lunacy commissioners and visiting justices of England.
The entire country is divided into four districts, each
having a head guardian and a physician, to whom is
intrusted the medical care of every inmate belonging
to the section. There are in addition one consulting
surgeon and one inspecting physician, resident in the
infirmary at Gheel, for the whole community. The
general government of the colony is vested in the hands
of eight persons, who dispense a code of laws especially
devised for it. The burgomaster of Gheel presides over
the managing committee, whose duties are to distribute
the patients among the different dwellings, watch over
their treatment, and to admit or discharge them. Of
late the divisional officers have the duty of selecting the
nourriciers, who are chosen, not hap-hazard like our
own, but for no other reason than the good of the
patient, and they are selected for him with a view to
his age, manners, language, and calling—in short, the
individual requirements of the lunatic are especially
considered. Even the style of household and family
arrangements is not thought too small a matter to take

into account, when the disposition of the lunatic is settled. The *nourriciers* themselves have a stimulus of the reward for their kind treatment, in the shape of a more remunerative patient, and they also have honorary rewards distributed with great ceremony for their kindness and intelligence; on the other hand, in case of any neglect, the patient is instantly removed, a punishment which is generally effectual in preventing a neglect of duty. It is said that the *nourriciers* have acquired through ages a traditional aptitude for the intelligent treatment of patients. This may seem a strange assertion, but we see no reason why qualities of this nature may not as well be transmitted—at least, if Dr. Darwin's facts are to be depended upon—as any others."

A later writer upon this remarkable colony, Dr. Edmund Neuschler, says :—

" At the hearth and at the table, so also in the stable and the field, and at the most various occupations, the working patient is the companion of his *nourricier*. At the time of my visit, attention was universally directed to the potato harvest; and I saw the liveliest activity out of doors, both among sane and insane. *The constant companionship permits the most natural and unconstrained supervision of the patient.* It does not annoy him, and it is hardly to be observed, as the *nourricier* does not stand over him like an idle spectator or a keeper, but is apparently engrossed in his own work. Often, indeed, if the patient is trustworthy, he goes alone to the field, or is accompanied only by a child; and it has never happened that the latter has been injured by his companion."

It cannot be expected that no restraint is used, considering that our system of non-restraint is nowhere received abroad; but it is worthy of notice, that with this free-air system of almost perfect domestic treatment, the number of persons in restraint, and that of a light kind—consisting mostly of an anklet—is less than is to be found in many of the closed asylums of France. Even these restraints—used mostly to prevent escape in a perfectly open country — are becoming milder every day, and the present chief physician, Dr. Bulkens, is in hopes of getting rid of them altogether. The remuneration to the *nourricier* is small indeed, compared with the sum allowed to patients' friends in England when they are permitted to go out on trial—namely from 65 to 85 centimes daily, out of which, besides expenses of clothing, 12 francs are annually deducted for medical attendance.

Ever since the existence of this singular community has been made known to the psychological world in England, its teaching has made the most profound impression upon it. It was not to be expected that our own superintendents of asylums, saturated with a vicious spirit of routine which they unhappily term experience, would at once acknowledge the value of a plan so antagonistic to their own interest and to our own asylum practice, which they have been led to imagine as perfection itself; but in the discussions that are continually taking place on the advisability of a further extension of the non-restraint system, Gheel is continually cropping up like a ghost that cannot be laid. Insignificant objections, futile nibblings at details, the usual outcries of small minds on the impossibility of *our* learning anything from a benighted remnant of a remote age, are heard from time to time; meanwhile, practically we are beginning to talk with approval of " the associated cottage system." But a moment's consideration shows that this plan, referred to by the Commissioners, is an inversion of the plan of the Gheel community. In the latter the hospital is a mere atom compared with the wide extent of the surrounding country, which is the real trial-ground and true fold and asylum of the patients. The asylum building is not even visible, and never throws a dismal shadow on the surrounding free ground, whilst our associated cottages are oppressed with the magnitude of the gloomy walls that overshadow them. The patients in them, whilst allowed this slight tether, feel that the attendants under whose care they remain, bring day by day the stifling asylum atmosphere with them, and all the associations of the dreary house of detention. And if these cottages thus overshadowed are sought after, as we know they are with delight by the patients, what a light the fact throws in the direction of Gheel!

Indeed it is in this direction that nearly every eminent authority in psychological medicine is inclined to tread. " Family life " is the new watchword that is being uttered by the best teachers on Mental Pathology throughout Europe. The family life mainly surrounds the woman; she it is who is its perpetual centre—from her flow all the affections and the feelings; we can therefore fully understand the reason that in the colony of Gheel it is the housewife that mainly takes charge of the patients. Dr. Brierre de Boismont, whose eminent authority is worthy of all attention, dwells particularly upon the merit of the feminine influence in the treatment of the insane.

"The character of man," he says, "cannot bend itself to this kind of slavery. The attempt to do so is indeed most distressing, as one must listen to the same complaints, the same pains, and the same demands. These repetitions last for hours, sometimes for days. They are mingled with disagreeable remarks, irritating words, insulting reflections, and even the infliction of bodily injuries, and very often accompanied by lying slander and calumny. The character of women accommodates itself better to these incessant annoyances."

Those only who are intimate with the insane know the value of these reflections; and not only may we add our own testimony to the value of these words, but we may also observe that the influence of children is incalculable for good. In the artless ways of the little ones there is nothing that irritates or alarms. The insane are rarely suspicious of a child's motives; they will follow their directions, join in their amusements, submit to their demands with a simple faith that is remarkable, considering the fear they too commonly entertain for the motives of adults. We give our implicit belief to the statement that in Gheel a child has never been known to have been injured by any of the male patients.

Dr. Maudsley, than whom we can mention no higher name among our own psychological physicians, has wisely prophesied "that the true treatment of the insane lies in a still further increase of their liberty;" and in doing so he is but liberally carrying out the forecasts of his father-in-law, Dr. Conolly; and Dr. Lockhart Robertson, the Chancery Lunatic Visitor, has practically endorsed the same doctrine in a letter lately written to the editor of the *Lancet*, where, speaking of the benefit of placing better-class patients in the houses of medical men as private patients, he says :—" The improved treatment of the chronic insane lies in this direction—in removing them when possible from the weary imprisonment of asylum surroundings, and in placing them amid the healthier influences of home life." "Many chronic insane," writes Dr. Maudsley, "incurable and harmless, will then be allowed to spend the remaining days of their sorrowful pilgrimage in private families, having the comforts of family life, and the *priceless blessing of the utmost freedom* that is compatible with their proper care." If this can be truly said of better-class patients, such as are to be found in private asylums, we cannot by any stretch of reasoning see why the same humane advice should not be extended to the insane pauper. The Sussex County Asylum, over which Dr. Robertson once so skilfully presided, admirably conducted as it is according to the present ideas of asylum management, can by no means compare with any well-conducted private asylum in the homelike character of its surroundings, or in the domestic nature of its treatment: hence we must claim him as an advocate for the domestic treatment of the pauper lunatic. We know it is asserted that middle-class life can furnish more appropriate accommodation to private patients than could the lower class for asylum patients ; but we hold this to be a wholly gratuitous assumption. Does any one doubt that if a premium of twelve shillings a week were offered by advertisement for the care of harmless pauper lunatics, that adequate accommodation would not be offered in abundance ? We think there can be but one answer ; and yet twelve shillings is much less than the actual cost per head of asylum patients. On the average the weekly estimate is about nine shillings, but this sum excludes the original building charge or house-rent. Considering the magnificent scale on which asylums are built, and the quantity of land they stand upon, an additional five shillings per head on this account must be added (the sum in reality is much more), yielding a present cost of (say) fourteen shillings for every pauper-patient in these establishments. Why should we persist in keeping these chronic cases against their will, and at such an expense, when, with more liberty and happiness, they may be maintained at a far less cost, and at the same time free the asylum of the beds they occupy for immediate and curable cases?

We are led by the result of these figures to consider the system adopted for pauper-patients in Scotland, the only plan that can be compared with that of Gheel. Their suitable cases are distributed among their friends and in licensed houses. The Scottish Commissioners report that this plan, which relieves the asylums of all patients that would otherwise inevitably tend thither, and removes those that would otherwise cumber the wards, works very well; and it good health is a criterion of good treatment, the Scotch pauper-lunatics so disposed of may be considered to enjoy a most unexceptional position, inasmuch as the mortality among them is lower than we find recorded among our own insane poor. Attempts have been made to depreciate this "Gheel of the North," as it has not inappropriately been termed; but we fail to find any reason for this disingenuous attempt. In 1866 there were 1,588 pauper-patients thus disposed of—75·5 per cent. with relatives and friends; 21·1 per cent. as single; and 3·4 per cent. to persons who have the Board of Lunacy licence to receive either one, two, or three patients under one roof. This arrangement appears to be an exact copy of the Gheel arrangements. Like the Gheelois, they are under the control of the Scotch Lunacy Board, and they are visited by the Commissioners at stated times, who have the power to remove any patient to an asylum, or otherwise as may be thought desirable. We gather from the report of Dr. Mitchell, one of the deputy-inspectors, whose duty it is to visit them, the following observations :—

"They will find more to interest them in the everyday occupations of a cottage life than they could in any large establishment. What goes on there, and what they see there, comes more easily within their comprehension and interest, *and they have a pleasure in feeling that they have some little share in it all, and that personality is not lost.* Their occupations and amusements may be more commonplace than in asylums, but they are not necessarily less useful on that account. The cottage kitchen is an ever-busy, shifting scene, and it would not be easy to manage a tranquil pauper-patient passing from acute disease into incurable imbecility, more favourably situated than at its fireside, where the surroundings are natural, and the influences are healthy."

It would be impossible to deny this statement with truth. The picture addresses itself to any unprejudiced mind as unexaggerated and life-like. It is charged by the asylum advocates with being drawn with a *coulcur-de-rose* tint, but we can see no sign of false colouring; neither is there any reason to call in question the strict veracity of Dr. Mitchell's statement. The only point in which we should feel inclined to differ from him would be his assertion that "such surroundings" are more applicable to the fatuous and idiotic, or mindless persons: all classes not dangerous would be equally benefited by such a family system. The English Commissioners " have their reasons for doubting whether the system could adequately be extended so as to afford any material relief to the county asylums : " giving no other explanation of this reason than that 6,600 insane paupers so reside with friends in England; but this can be no bar to a further extension of the system under much better control. The country is large enough to support ten times 6,500 pauper-lunatics, if means were taken to establish such a system. No doubt, sixpence a day, the Scotch allowance for such patients, is not sufficient; but, as we have before stated, twelve or fourteen shillings a week would be amply sufficient. The Commissioners, overworked as they are at present, would, we admit, be totally unable to undertake the very necessary work of supervising such a crowd of patients as would be thus accommodated; but this objection could be remedied by an increase in their numbers. They may trebled with advantage; or, if this plan would be too costly, the work of supervision may be undertaken by the union medical officers at stated times in the year.

It has been proposed that the supervision of such cases of chronic and harmless lunatics, thus boarded out, should be entrusted to the superintendents of asylums. This plan would occupy the time of that official, which would be much better employed with the acute cases in the asylum. Very little, if any, medical care is required for those poor people who are beyond the physician's art. Moreover, the plan of entrusting their supervision to the asylum superintendents would, we believe, be injurious in two ways. In the first place, in order to save time, there would be a tendency to lodge such boarders as near as possible to the asylum,—to make a colony close to its doors. Now this may very well satisfy the superintendent, who would wish to retain his dominion, and to maintain a certain kind of modified restraint upon the actions of the patients; but we contend it would be an unnecessary encroachment upon their liberty, and therefore injurious to them mentally, inasmuch as they would still feel themselves to be under the depressing influence of the prison from which they had been liberated; they would be a kind of ticket-of-leave lunatic, and would partake of the ticket-of-leave man's dreads and suspicions. Of course, where convalescent cases were thus lodged out of the asylum, as near a contiguity to it as possible would be advantageous, for the sake of the physician's constant attendance; but the chronic lunatic may very well dispense with his visits.

Supervision by a paid staff of inspectors we hold to be indispensable in such a free-air system; and we believe it to be the most practicable and the most advantageous, both for the sake of the lunatic and for the sake of the asylum itself. The visitation of private patients at present is a mere delusion, once a year being the average amount of visits paid to them. In the case of pauper boarders they would demand more careful supervision than even the better class of patients : hence a large increase of the inspectors is indispensable. It would require time to get such a system into working order; but it would, when once established, be so elastic, that no new rules or regulations would be demanded. The office of such inspectors should not only be that of supervision, but they should also have the duty of distributing the patients. We do not think that the cottage system, pure and simple, is the best adapted to the class of patients such as are found in the neighbourhood of important cities; for instance, Colney Hatch and Hanwell number among their inmates a large majority of town-bred lunatics. These would not necessarily be benefited by being placed in cottages in rural districts. Their habits and associations are all connected with town life. The country lunatics, again, would live more happily amid the fields, and in the midst of rural occupations among which they may take a part. Following the admirable example of Gheel, the inspector should have the power of placing out the pauper-lunatics in such houses and situations as would be best fitted for them. The peculiarities of each case should be considered as far as possible, and the person taking charge of it should be the most suitable.

There is no reason why the pauper-lunatic boarder should in this respect be treated worse than private patients in private houses. Indeed, what we ask for them is a perfectly similar free-air treatment to that granted to the quiet chronic cases among the better classes. London is full of certified patients, many of whom mix with the general population; but we never hear of offences committed by them, neither should we if harmless pauper cases were distributed among the population.

If thirty per cent., and this we believe to

be below the real number that could with advantage be withdrawn from our asylums, were thus boarded out in private families, all the difficulties with respect to finding beds for acute cases would at once vanish, and the perplexing problem which is at the present moment troubling asylum physicians, commissioners in lunacy, visiting magistrates, and the taxpayers, would be solved. The existing establishments would present vacant wards, instead of being crowded to suffocation, and civilization would no longer be outraged as it now is by the daily refusal to admit urgent cases. According to the last report of the Commissioners in Lunacy, just issued, no less than 661 applications for admission had been refused at Hanwell, and 562 into Colney Hatch, in less than twelve months!

But it is not sufficient to remove these chronic cases from the county asylums, we must prevent fresh ones getting in, which would speedily happen if some change were not made in the terms of their admission. Harmless cases of long standing must be made inadmissible, just as they are at St. Luke's. Unless the door is shut to cases of this kind, which are beyond hope of cure, it would be impossible to free asylums of the dead weight that would inevitably again oppress them. They may be admitted for a short time, in doubtful cases, but immediately the physician has ascertained that they are past cure, they should at once be drafted out into private houses and keeping.

And here, we may ask, may we not take some steps to arrest the disease before it has become fully developed? It is well known that the curability of the disease depends upon its being treated early. But how is early treatment to be secured for the poor? It has struck many thoughtful minds, that one crying evil of the treatment of insanity is the fact that it is made a special science, apart from the ordinary range of general medicine. By the general practitioner it is looked upon as something out of the way of his regular duties. The family doctor has not been accustomed to consider such cases, and when brought before him, he refers them to a special authority, as something mysterious that ordinary medicine cannot touch. The approach of an attack is either unobserved, or treated simply as low spirits, or the result of indigestion; possibly the practitioner has never seen a case of mental disease, is totally unskilled in the symptoms, which, to a trained mind, would have given forewarnings of an impending attack. This is a fatal blot in our medical teaching. Insanity is as much a bodily disease as gout or rheumatism. The insane action or idea as surely springs from a morbid derangement in the brain structure, as a bilious attack springs from a morbid condition of the liver. There is no mystery about it; it is a mental manifestation arising from a physical cause, and should form as necessary a branch of medical study as chest or heart disease. We believe ourselves that this separation of one organ, and that the highest, the brain, from general medical study, is the most fruitful cause of incipient insanity being suffered to degenerate into confirmed lunacy. The sentinel who is at every man's door, be he rich or poor—the general practitioner,—is the one who should be able to foresee the approach of an attack. But he has never studied, or has the slightest possible knowledge of, psychological medicine; the danger goes on from day to day, the chance of averting the evil is lost, and when the patient has become an outrageous lunatic, he is taken to a " mad doctor,"—that is, if he has the means to pay his fees, if not he is allowed to linger on, making his home miserable, and sinking every day into deeper disease, when he is taken to the asylum.

The loss to the community by reason of this defect in the knowledge of the general practitioner is not the only evil of this separation of psychological medicine from general medicine. The error which underlies all special study and experience, even if it makes the vision keener in a limited area, is far more serious where mental afflictions are concerned than in other diseases. A surgeon may with advantage devote himself to particular manipulative arts. A man who is drawing teeth all day makes a far better dentist than a general practitioner. The operation of lithotomy requires special skill, which practice alone can secure. But to treat mental disease properly, not only the condition of the brain, but of the whole body, must be taken into account, as in all cases madness arises from morbid bodily conditions, some of which the specialist overlooks, or rather he is so engaged in looking for one thing, that he overlooks another which may be of equal or greater importance. Of course, there will always be physicians eminent in mental disease, leading men whose genius in their own department overrides all other shortcomings, but these will necessarily be few. Otherwise we are convinced that, for the good of general medicine, this particular study, dealing as it does with so many complex problems, should be merged in the general routine of medical practice. If insanity were treated as a purely physical disease, like any other nervous disorder, it would lose half of the dread which at present surrounds it; it would no longer be hidden like a crime, and the patient himself would not feel the misery of being avoided and distrusted,—one of the most annoying things that meet the convalescent, and often the cause of the distrust he himself evinces. Moreover, there would be no fear of positive injustice being done to the poor man, such as the decision of the late Mr. Tidd Pratt threatens to inflict upon all members of Friendly Societies who may happen to become insane. This gentleman, apparently taking the old priestly idea of insanity, that it is a spiritual disease, and therefore not within the range of usual physical maladies or infirmities for which these societies give aid in the shape of weekly sick allowances, refused to certify the rules of any society that proposed to give such aid; indeed, in more than one instance, sick allowances have been refused to members thus afflicted with the most pauperising of all diseases. When the universities and other licensing bodies demand a knowledge of mental disease from all graduates in medicine, insanity will meet with an important check to its future progress.

But the first step towards a proper utilization of our present system of treating mental disease in our public asylums is to disgorge them of the cases that clog their action. A fatal torpor seems at the present moment to affect all parties interested in this necessary reform. The Commissioners, the medical superintendents, the visiting magistrates, and the taxpayers, whilst admitting the evil, seem to have lost all power to make a change. Meantime, as the asylums are becoming monstrous by gradual accretion, a still more fatal obstacle to the further application of the principle of non-restraint is going on. The amount of capital sunk in the costly palaces of the insane is becoming a growing impediment. So much money sunk creates a conservatism in their builders the county magistrates, which resists change; and, moreover, vested interests are growing up, which unconsciously warp the minds of the medical superintendents, as any great or radical change in the treatment of the insane would, they imagine, endanger their present position,—an idea which is, of course, erroneous, inasmuch as in no case can the treatment of acute disease pass into other hands. Hence the strange and futile objections that we see daily urged against a greater freedom in the treatment of the lunatic; but that a sweeping change in that direction is one of the inevitable reforms we feel blowing towards us in the breath of every angry discussion among practical psychologists on this matter, is but too obvious. As we see wing after wing spreading, and story after story ascending, in every asylum throughout the country, we are reminded of the overgrown monastic system, which entangled so many interests and seemed so powerful that it could defy all change, but for that very reason toppled and fell by its own weight, never to be renewed. Asylum life may not come to so sudden an end, but the longer its present unnatural and oppressive system, as regards the greater number of its inmates, is maintained, the greater will be the revolution when at last it arrives.

THE TRAINING OF IMBECILE CHILDREN.

THERE are upwards of fifteen thousand imbeciles in Great Britain at the present moment, the greater portion of whom are not only incapable of helping to do their share of the work of the world, but who absolutely detract from it, inasmuch as many of them require to be watched, fed, and dressed by those who are not afflicted as they are. Fifteen thousand is a little army, the majority of whom are a burden upon the parish rates. Generally belonging to the poorer classes, who cannot afford to maintain them, they are transferred to the care of the various workhouses throughout the country, where they pass their lives without gaining an idea, and die like the beasts of the field, only a little more helpless. In foreign countries the idiot and imbecile population has not been allowed to perish without an attempt to improve it. Although the larger proportion of imbeciles is absolutely incapable of any culture, a very decent per-centage—at least thirty-five—have been found capable of making some intellectual progress, at least sufficient to raise them to a position superior to the level of mere animals, and in many cases to cultivate their dim intellects sufficiently to enable them to help themselves. In England we have been somewhat behind-hand in these philanthropic efforts. The State has done nothing towards rescuing even the number that are capable of improvement out of the Slough of Despond in which they wallow in the pauper asylums throughout the country, and the work has been left, like so many other purely national objects, to be carried out by private charity.

Earlswood Asylum and Essex Hall are the only institutions in this country as yet which are employed in the labour of training the poor imbecile, so that he may at least not be a burden to others, and in many cases that he may become a very tolerable artisan or workwoman. We are all familiar with the building at Earlswood, as we pass it on our right hand at Redhill on the way to Brighton, even if we do not know the really Christian purpose for which it was built. The asylum, it may be stated, is not a mere receptacle for the helpless idiot—indeed, such are not eligible for admission within its walls—but only for those whose brains are sufficiently well formed to be capable of receiving instruction. Thus the institution is a hospital, and not a mere receptacle for effete humanity, which cannot be improved by human aid, and which is therefore best left to the protection of the union.

The visitor to Earlswood need not fear to meet with any of those repulsive objects scarcely in human form whom he is apt to associate with the word idiot or imbecile. He may, therefore, enter without fear, and witness one of the most interesting sights to be met with within the field of philanthropic labour. The majority of the children reared here are simply specimens of what is termed "arrested development." At a very early age the functions of the brain—at least in regard to its intellectual operations—appear to have stopped; hence we see a school full of grown-up boys and girls, sometimes of the age of eighteen or twenty, no more capable of taking care of themselves than children of three or four. They cannot use their hands in any ordinary operation; sometimes they do not know the way to eat with a knife and fork, and as a rule the new-comers are utterly incapable of dressing or undressing themselves. Their very actions are those of little children; their emotions and fears, their joys and sorrows, remind us most forcibly of those we witness in a nursery of little ones. Here is singular raw material for the intelligent physician to work upon, and well does the physician perform his task. Dr. Down, the resident medical man, had to begin his task from the very beginning. The work of the nursery, which is sufficient to make the little one of ordinary intelligence an adept in small personal matters and attention, has to be gone through with these bigger babies. The faculty of imitation, which is common to the monkey and to the child, is the great instrument by which these poor little ones are taught to exercise their senses, and to acquire the ordinary habits of civilized beings. They are grouped in classes, a few of those already instructed being mixed with those who have to learn. Finger lessons are the first that are taught. Most of the children, for instance, on first admission, cannot button a button, tie a string, nor do the commonest act which requires any adroit manipulation of the digits : hence all this has to be learned. It is certainly an odd sight to see a group of girls all actively engaged in buttoning and unbuttoning their clothes, in pinning them, and in tying and untying strings. In a short time, by watching those who are instructed in these simple arts, they become adepts, and are able to dress themselves with perfect ease. Whilst witnessing the mechanical manner and the earnest expression with which these lessons were performed, we confess we were reminded of the performing monkey the Italian organ-grinder carries about with him, who sweeps with a broom, plays the drum, shoulders and lets off a musket, and does half a score of tricks with equal adroitness. The use of the limbs is taught in classes in the same manner, the very exercise of volition giving immense pleasure to these little ones. Even the soundest-brained children flag in their attention if kept confined to one task too long, and to these imbeciles the faculty of attention is doled out in the most infinitesimal quantities : hence the necessity of constantly changing their occupations.

In all imbeciles the powers of speech are very imperfect : this arises in many cases not from any deficiency of the organs of speech, but from an inability to place the tongue and lips in a proper position to articulate. This is remedied by associating the children together in what is termed the "bell-ringing lesson." They are taught to imitate the action of bell-ringing, and whilst in this manner exercising the organ of time, they are taught to sing in chorus some nursery song, such as " Ding, dong, bell," &c.

By this means the organs of speech are taught mobility, and in a short time that which they learn parrot-like they are able to repeat as an effort of will. By degrees those who entered the asylum incapable of any articulate speech, are taught by this method the use of their tongues.

To make workmen, however imperfect, out of the imbecile, there are a great many qualities beyond those we have mentioned necessary. They must have a tolerably accurate idea of form. This is a difficult thing to teach, but it is done by giving them lessons in fitting things together. Thus, one of the lessons is to place a number of square and oval pegs before the imbecile, and teach him to fit them into apertures of a corresponding character. This form of instruction is particularly useful to those who are intended to be taught carpentry. Various handicrafts are taught here, and therefore many preliminary lessons of a similar character have to be acquired before the boys are entrusted with tools. All the tailoring is done by the inmates, a regular teacher presiding, and instructing the more advanced in cutting out, sewing, and fitting. The workmen give the spectator, however, the idea of being boys at play, for they come up to him and eagerly show their work, as the youngest children show their drawings on a slate, delighted at the smallest praise. Besides the handicrafts, the boys are taught agricultural pursuits; they cultivate the garden, and feed the stock—the favourite occupation; they milk the cows, and do it well, too. Five years is the term those are admitted for who are upon the foundation, and by the time this period has expired very many of them are able to take part in the work of their homes, instead of being a mere encumbrance, often filthy and disgusting in their habits. The

The Training of Imbecile Children. 173

routine instruction of the asylum lifts all of these poor children, more or less, into the scale of rational beings, able to help themselves and others; but in some instances the most encouraging results are obtained. Thus, some of the lads are able to copy engravings in a surprising manner. It is done in a purely mechanical spirit, it is true, and this very fact affords a proof of the small intellectual merit of the merely copying capacity; but are there not many persons earning their bread in the world, and considered to be very clever, by means of the same limited powers? There is a lad here, however, who shows no mean constructive ability. Doctor Down discovered that he was fond of cutting out ships from the solid wood, and to encourage him had him taken down to Woolwich Dockyard, where he witnessed the building of some vessels of war. The hint was sufficient for him: henceforth he disdained to cut his model out of wood, but set to work to build it up after the manner of a regular shipwright. He even made preliminary working drawings of the different sections, no mean intellectual effort, and from these drawings constructed his ship in a most workmanlike manner. It is preserved as a trophy in the asylum of the skill of the trained imbecile. Fired with this success, he determined to make a model of the "Great Eastern," the most elaborate drawings for which we saw some time since laid down on the drawing-board, including some most mathematical-looking midship sections, which few boys of much sounder intellect would be capable of drawing, much less of working by. For some reason, of which we have not been informed, this new venture has never been proceeded with; but he has constructed numerous models, which were exhibited in the Paris

Exposition. Yet this young man appears to be infinitely below the level of many of the other inmates in general acquirements; he speaks with the utmost difficulty, and then but very imperfectly; and, when he can, he makes himself understood by a rude system of hieroglyphics, which he draws upon a board.

The girls, after they have conquered the preliminary difficulties of what we may term nursery instruction, are taught sewing and dressmaking, and have to take part in the household work, the greater portion of which they do under the direction of female trainers. It is a well-known fact that a far greater number of males are imbeciles than females, and all the worst cases of idiocy and imbecility belong to the former sex; but in Earlswood, at least, there are no show girls that come up to the show boys in their intellectual efforts.

We forgot to mention, among the latter, the boy with the wonderful memory, which he exercises in so surprising a manner, that if he had been of the outside world many persons would have thought him to be a remarkable prodigy. This lad goes through the History of England with the most extraordinary fluency, and with the utmost accuracy. But in some mysterious manner one link of facts seems to be so connected with the one following it, that when once he drops it his memory entirely fails him, and he is obliged to try back again until he picks it up once more. This is a proof of the purely mechanical nature of the process. Otherwise, he shows no quality above the average of the boys about him.

The value of associating imbeciles together, however great, as a means of teaching each other, was soon found by Dr. Down to have one serious drawback. The very fact of their being withdrawn from every-day association with the sane, deprived them of the power of self-reliance. They lived, in fact, in the narrow school of the asylum instead of the great school of the world, and it was accordingly found that when they went out they were as far as ever from being able to deal with the sane people from whose ways they had been withdrawn. Having their food and every other necessary found them every day, they lost the knowledge of the value of money; they were ignorant of the commonest machinery by which the great world is carried on. To meet this patent deficiency a new system of instruction had to be adopted, and for this Dr. Down took a leaf out of the playing instincts of the nursery, in the institution of "keeping shop." This is the most interesting sight, to our mind, of all the operations carried on in the asylum. A room is fitted up as a general shop, furnished with weights and scales, and all the paraphernalia of a real counter. At the back there are drawers, in which is to be found all the miscellaneous collection of articles to be met with in such establishments. The names of the articles are written outside the drawers, and everything is complete, with the exception of the salesmen. The rows of seats which are ranged along one end of the room are speedily crowded with the lads, eager for the fun. A shopman is called for, and a score of boys volunteer. One is picked out, and then the instructor calls for a customer, and, of course, crowds offer themselves, and at last a lad is selected; and the little drama, played with real earnestness, commences by the buyer marching into the shop, when the shopman is expected to salute him politely, and if he does not, he is reproved by the customer. We all know how strict children are when keeping shop; how they insist upon every act of the play being performed with the most undeviating regularity. It is exactly the same with these children of a larger growth but of a weaker brain. The play has, doubtless, been performed so many times that they have it all by heart;" but it is not the less interesting to watch it through its different stages. The real object of the little drama is to give the children an idea of the value of the articles they will be required to purchase in after life, and the use of the weights and scales. A regular higgling match forms the conclusion of the drama. An article is asked for, the weight demanded, and a regular calculation is gone into. The quarrel about the right change to be given is very animated, the whole audience giving advice, or prompting when the chief actors are at fault. The instructor is present all the time, keeps order, and sees that the transaction is carried on in a fair and proper manner. There can be no doubt of the real value of this lesson, taught under the guise of play, and the beauty of it is that the poor brains of the actors and audience are relieved rather than fatigued by the performance. These exercises are a preliminary to the sending the boys outside the walls of the asylum on small errands, such as with letters to the post, &c. In this way the errors inherent to a strict asylum life are corrected.

Sometimes they are treated to real amusements, such as Punch, the lime-light, and the galanty show; by these means they are made familiar with the various animals and scenes existing in real life. They are presented, in short, with little pictures of the world, and of the relations different objects have to each other.

There are all classes of patients in the asylum, but they are by no means mingled. It is, in fact, in one respect like a hotel, where different accommodation can be obtained by paying for it. The first class of patients have their own private apartments and nurses, in a part of the building quite distinct from the other portion of the establishment; the next class, paying fees of thirty-five and fifty guineas, can associate with those who are elected on the foundation. All the patients, however, go through the same system of training—not station, but intelligence forming the method of classification adopted. This question of rank does not in the slightest degree affect the children themselves, who are not cognisant of the differences of social standing; the poor man's son and the child of the rich one, are going through their lessons side by side, in the happiest and most contented manner. What a pity it is that people with stronger brains cannot do the like!

A comparison of the condition of many of the imbeciles upon and some time after admission, is a measure of the real value of the training they have received. Those who know anything of the ways of imbeciles, must have noted their filthy habits, and their tendency to tear everything they can put their hands upon, their own clothes of course included. This tendency is utilized in a very ingenious manner; a fabric is put into their hands that requires picking to pieces, such, for example, as oakum. Of course they exercise their destructive tendencies upon it, but are surprised to find that instead of being scolded as they used to be, they receive kind words and thanks for their obedience. Thus the very first lesson they learn is that their new teachers are not inclined to kick and cuff them about, but to praise and speak kindly to them. The effect upon the newcomers, low as they may be in the scale of intelligence, is very marked, and it pretty generally secures at once their better feelings and their willing obedience.

We are not aware what the methods of instruction are at the only other asylum for imbeciles at Essex Hall, but we doubt if they can be better adapted to the purpose than those Dr. Down has, with so much intelligence and real knowledge of the true mental condition of these poor creatures, put in practice.

It is just possible that thirty-five per cent. out of the fifteen thousand pauper imbeciles in the country, may not be the limit to which a certain amount of instruction may be conveyed. We do not know, in fact, until we try, what depths of mental deficiency may not be improved by proper treatment, and it seems to us to be the duty of the State to test every imbecile before it throws him aside into the workhouse, as a piece of waste humanity, of no account in the great scheme. If the majority of these poor creatures were of the better class, there would be no fear but that human skill would be brought into play to lift them from the mere animal life to which the mass of them are doomed, into the higher level of self-reliant beings—of a very inferior order it is true, but still vastly superior to the drivelling human abortions that now crowd our union workhouses.

We reiterate our praise of this institution for the training of the pauper class of children, but we think the method of mixing the better class with them is an error; in as far as their manners and habits are concerned, children are but too apt to follow the habits of their fellows, and even in their imbecility the habits of the lower class are far more repulsive than of the superior order.

Again, unless there is a necessity, in a pecuniary point of view, we hold it is a mistake to associate poor imbeciles : they require rather the widest possible mixture with the sane members of society, just as lunatics do where it can be done with safety. If the poor could be distributed among the working-class families, the teaching would be far more real than the playing at life in these imbecile schools ; but we could not, we fear, depend upon the food and clothing they would receive from persons in their own station in life. This objection, however, in no manner obtains with the better class ; they should be with the families of medical men, where they would get a thousand times better training among the children of the household than could be obtained in any institution, where at best their teaching is of a make-believe character, in an atmosphere sodden with idiocy, and, therefore, frightfully intensified.

There is nothing that is put forth in Dr. Down's scheme of tuition which could not be put in practice in the family of a medical man receiving such unfortunate specimens of waste humanity for improvement— nothing, as we have said, but the deadly element of concentration, which is the one great thing to be avoided, where possible.

The object of recounting the valuable educational advantages which pauper idiots are receiving, is to show that the same human material in the upper and middle class of life are at present very much worse off than their inferiors in the social scale. That there is no necessity for this inferiority must be obvious to any medical reader. The teaching power of the nursery must be far superior to that of the asylum, inasmuch as in every nursery, however low it may be in intelligence, some members of the family must be in advance of its idiotic member, and, therefore, more capable of imparting their knowledge to others. Where, as in some medical man's family, the children are of the normal standard of intelligence, the teaching power is proportionately greater. Although, therefore, we think every praise is due to Dr. Down for his invaluable utilization of the power of imitation in the education of children of the lowest intelligence, it must not be conceived that there is any especial value in the teaching power of an institution containing a large number of children. Here again, as in the case of chronic lunatics, the value of mixing them with healthier brains than their own is of the utmost importance—in fact, by Dr. Down's own admission his scheme breaks down at the very point where the real education of the child, as respects his intercourse with the outer world, begins. Our own opinion is, that no idiot of the better class is incapable of development in an educational point of view to a degree which would render him capable of attending to his own immediate wants, and it is only where the individual falls short of this that the repulsive nature of his case is apparent. Of course, even a slight modicum of education can only be acquired by these unfortunates in very early youth; we cannot therefore impress upon our readers the necessity of giving their earliest attention to the culture of this waste material of humanity. If we may say so much of the very lowest forms of idiocy, with respect to the highest forms the value of intelligent training cannot be too highly estimated.

BRAIN ENIGMAS.

ECCENTRICITY is the mild term under which we favourably disguise a more painful affection—that is, if we are to consider that any prolonged departure from the normal sensations of mankind may be said to constitute disease. Many mental eccentricities are only the forerunners of serious mental failure, and are, indeed, recognized as such by the more intelligent and honest physicians.

In many of the more terrible lesions of the brain, resulting in entire loss of intellectual power, the very smallest symptoms are often indicative of the mischief that is about to intervene. The inability to grasp a stick, the continual numbness of a finger, the loss of memory in small matters, are often indications of serious cerebral disturbance. Dr. Graves, a Dublin physician, tells us that "an Irish farmer, in consequence of a paralytic fit, lost the power of remembering noun-substantives and proper names; the extraordinary thing was, he could remember the initial letter of the words he wished to say, but no more. In order to meet this singular difficulty, he constructed a dictionary, including the names of the articles he was in the habit of calling for, and also for the names of his family and servants. Thus, when he wished to speak of any of those persons or things, he turned to the initial letter, and as long as he kept his eye upon the word or name, he could pronounce it, but his power to do so was lost immediately he missed the place." We do not know the ultimate history of this case, but in all probability this extraordinary symptom was as fatal as it was singular.

Even the use of a wrong initial letter, if persisted in, is a symptom of mental disturbance. If you hear a man continually saying "puc" for cup, "gum" for mug, or "etulf" for flute without being aware of it, you may be pretty sure that his brain is affected. It often happens that such mistakes are made by the best of us in ordinary conversation, but we instantly become aware of the error and correct ourselves, and this is just the difference between the sound and unsound brain.

In some curious cases of mental disease persons have wholly forgotten acquired languages and reverted to their native tongue; on the other hand it is recorded that Dr. Johnson, when he was dying, attempted in vain to repeat the Lord's Prayer in English, but did so in Latin. In the same manner, certain events will slip out of the memory altogether.

Dr. Pritchard tells us an anecdote which proves that the brain stands still for years upon the invasion of disease, and when the attack has passed, can take up the recollection of an action, just at the point at which it had left off.

A farmer of New England, whilst enclosing a piece of land, happened, when he had finished his day's work, to put the beetle and wedges which he had used for splitting the timber, in the hollow of a tree. That night he was seized with a mental attack which prostrated his mind for many years. At length, however, his senses were suddenly restored, when the first question he asked was whether his sons had brought in the beetle. They replied that they could not find it, fearful of bringing back a recurrence of the attack; upon which the old man got up, went straight to the hollow tree, and brought back the wedges and the ring of the beetle, the wood-work itself having rotted away.

Sometimes during mental illness a patient will forget all the early events of his life, and upon a recovery lose knowledge of all late events, and recur to those of which he was previously oblivious. Samuel Rogers, the poet, towards the latter end of his long life presented peculiarities of memory very similar to those we have related of undoubtedly diseased brains. In Earl Russell's life of the poet, he says: "In his ninetieth year his memory began to fail him in a manner that was painful to his friends. He was no longer able to relate his shortest stories, or welcome his constant companions with his usual complimentary expressions. He began to forget familiar faces, and at last forgot that he had ever been a poet." The "Edinburgh Review" tells us that "although his impressions of long-past events were as fresh as ever, he forgot the names of his relations whilst they were sitting with him."

Pressure upon the brain by pieces of bone forced inwards by some accident, sometimes produces a complete death of the sentient being for the time, and a restoration of the mental power results upon the removal of the cause.

A singular example of loss of mental life occurred during the battle of the Nile, when a captain was struck on the head whilst in the act of giving orders. A portion of the skull was driven in upon the brain, and the officer at once became unconscious. In this condition he was taken home and removed to Greenwich Hospital, where he remained for fifteen months, living the life of an inanimate vegetable. Upon the operation of trephining being performed, however, his consciousness immediately returned; he rose up in his bed, and, in a loud voice, finished giving the order he was issuing when he was struck down.

One can understand the recovery of a man's wits, in consequence of a removal of a pressure upon the brain, but unprofessional people can scarcely understand a born idiot turned into a man of ability by a blow upon the head. We have the fact, however, upon the high testimony of Dr. Pritchard. "I have been informed," he says, "on good authority, that there was some time since a family consisting of three boys, who were all considered idiots. One of them received a severe injury upon the head; from that time his faculties began to brighten, and he is now a man of good talents and practises as barrister." And we may state that this is by no means a singular instance, recorded in volumes of psychology, of the brightening of the brains of idiots by reason of hard knocks.

One of the earliest indications of softening of the brain is often found to be a paralysis of the muscles of the face. The eyelid will drop, or the mouth will appear slightly drawn aside. A paralysis of the nerves of sensation sometimes leads to painful mistakes. A case is recorded of a gentleman who scolded his servant for having brought him a broken wine-glass at dinner. The servant, after looking at it, said it was a sound one; the master again put it to his lips, and again said there was a piece out of it. Ultimately it turned out, however, that the master was seized with paralysis of the nerves of sensation of one side of the lip, and not feeling the glass in that spot, of course concluded that it was broken.

The commonest symptom of approaching paralysis is a numbness of fingers, sometimes only of one finger, or a want of power in the grasping action of the hand. We do not allude to partial attacks, which often happen to the healthiest, but to persistent loss of either power, or sensation. The gait of a person is often highly indicative of some slowly proceeding cerebral mischief. A man so suffering will either fail to plant his foot down decidedly on the ground, or his foot will slip aside, or he will drag his leg. Sometimes the motions of a person suffering from incipient brain disease, are very like those of a drunken man, and it has often happened that an individual has been charged with having thus disgraced himself, when in reality he was deserving of pity and commiseration, in consequence of being on the verge of an attack of paralysis.

In the general paralysis of the insane—that most terrible of all diseases—the premonitory symptoms are of the most singular character. The individual so suffering becomes boastful, and extravagant in all his notions. However straitened may be his circumstances, he believes he is overflowing in riches; everything is successful with him; he can do everything better than any one else. In many instances these extravagances are the only premonitory symptoms of softening of the brain — that terrible malady which gradually saps every power of mind and body, and reduces the poor sufferer to a living death.

Those who are not acquainted with the various and extraordinary phases of the mind, and derangements of the moral faculties which herald an attack of insanity, are apt to denounce that plea put forth sometimes on the behalf of prisoners at the bar. There is nothing more certain, however, to the medical man versed in mental diseases, than that the perversion of the moral sense is, for a time, in many cases the only recognisable sign of an unsound mind.

There is one well-known form of insanity, kleptomania, which makes itself known by petty pilfering. This has become a sort of joke and by-word, and although it is perhaps difficult sometimes to recognise it in criminal cases, it undoubtedly exists. A letter-carrier, for instance, having been suspected of stealing letters, his house was searched, and thousands of letters, many of them two or three years old, were discovered hidden away. When a man or woman of well-known wealth is found out, committing these petty thefts, where there can be no reason for the act, the psychological physician, upon inquiry, soon finds out that the mere act of theft is only one of many symptoms which proclaim the person to be of unsound mind. A case is related which bears upon this point exactly, and which proves that in this form of insanity a very different treatment is dealt out to the rich in comparison to that dealt to the poor. A lady of good family and of affluent circumstances, accompanied by her maid, entered the shop of a fashionable jeweller at the West-end of London. The lady as well as the other members of her family, were in the habit of dealing with the tradesman referred to. After examining many articles of jewellery, she left the shop without purchasing anything.

Some time after her arrival at home, the master of the shop called at the house and requested an interview with the husband of the lady. This was at once complied with. He then informed him that his wife had been to his shop, and had, as he suspected, abstracted a valuable diamond bracelet.

The matter was immediately investigated, and the suspicion of the tradesman proved to be correct. This unhappy episode suggested an investigation, and, to the astonishment of her husband and all the members of her family, a number of diamond rings, valuable bracelets, gold chains, &c., were found in her possession of which no account could be given. This lady ultimately went completely out of her mind.

We may state that there are several members of wealthy and noble families at the West-end who are known to labour under this form of mental disease, and whose failings are so well known to the tradesmen with whom they deal, that they are allowed to take things, as they fancy, unobserved, but an account of them being immediately sent to the family, they are either returned or paid for.

Such persons, although thus painfully afflicted in a moral sense, will go on for years without showing any intellectual failing, or, indeed, without showing any particular oddity of manner or departure from their ordinary habits of thought, although latent disease is, no doubt, present.

In such cases the brain, after death, very rarely shows any perceptible departure from a healthy condition. Science is not yet subtle enough to detect the changes that must have taken place in its structure in order to bring about such a lesion of the moral sense. Indeed, it may be said that the cerebral matter of many confirmed lunatics exhibits no appreciable change from the healthy brain; on the other hand, this organ has often been discovered after death to be most extensively diseased, without any sign of mental disturbance having been exhibited by the person during life. Dr. O'Halloran mentions a very singular case in which a man, having suffered an injury on the head which caused the suppuration of the skull, nearly one-half the brain was discharged through the opening; nevertheless, this man, we are told, preserved his intellectual faculties till he died.

Soldiers have been known, again, to carry bullets in their brains without being much inconvenienced thereby; and, what is more, having passed through operations for their extraction, which resulted in a further injury to the cerebral mass, did not thereby suffer permanent loss in their mental power. Possibly the reader may doubt, in such cases, whether there was any mental power to lose, but it by no means requires the possession of intelligence to form the ground for insanity, the development of which is compatible with a very low condition of the mental faculties. On the other hand, the clearest and highest mental powers have been maintained intact, under circumstances which would have been thought fatal to the preservation of any of the senses. Dr. Wollaston, for instance, was afflicted from an early age with an abnormal cerebral growth, which increased with his years, and at last attained such a size that it pressed upon the cavities of the brain and produced paralysis of one side of the body. Nevertheless, his reasoning power remained perfect, and when he was struck mute from the same cause he was still able to convince his friends, by the performance of most abstruse calculations, that his brain was clear up to the last moments of his life. Nothing is more puzzling to the pathologist than these anomalous cases. When brain-disease is impending the special senses are often most acutely active. Dr. Elliotson relates a case in which a patient, previous to suffering an attack of hemiplegia, heard the slightest sound at the bottom of the house, and the ticking of a watch placed on a distant table, when the sound was quite inaudible to any other person. The senses of smell and touch become exaggerated in the same way. Everything will feel cold to the touch, sometimes gritty or greasy. We ourselves know a case in which a young man was always washing his hands under the idea—quite an imaginary one—that they were greasy. He died not long afterwards of well-pronounced brain-disease. Prout mentions the case of a man confined in the Bicêtre, who in the depth of winter, when the thermometer stood at twenty, twenty-five, and even thirty degrees below freezing-point, had such a sensation of heat in his system, that he could not bear a single blanket, but remained seated all night on the frozen pavement of his cell, and scarcely was the door opened in the morning, when he ran out in his shirt and applied quantities of snow to his chest, and allowed it to melt with delight like that experienced by persons when breathing cool air in the dog-days."

In others, the sense of smell is augmented in an extraordinary degree, and also perverted. Some persons on the eve of an attack of insanity will protest that their insides are putrid, and that the smell arising therefrom is intolerable to them; the touch, in the same manner, is often vitiated.

Dr. Simpson of Edinburgh, in describing the incipient symptoms of general paralysis, says that patients complain of their fingers feeling like sausages. A tailor who died of this disease, for twelve months previous to any suspicion of this malady coming on, could not feel the tips of his fingers, and consequently could not use his needle. It is a well-known fact, that some lunatics, on the approach of a paroxysm, feel on the other hand an irritation about the ends of the fingers which leads them to bite their nails.

The warnings, in short, of the approach of mental disease are very numerous, and would be very valuable if they were known in time; but to the patients themselves they always appear trivial. Indeed the symptoms are very similar to those that we experience in health, the real difference consisting in the persistency with which they remain, where real disease of the brain is involved. Thus the change, without any apparent cause, which takes place in the habits of persons, may be looked upon as singularly indicative of mental disturbance. It has been well remarked by a popular physician—Dr. Andrew Combe—that it is the prolonged departure, without any adequate external cause, from the state of feeling and mode of thinking usual to the individual when in health, which is the true feature of disorder in mind; and the degree in which this disorder ought to be held to constitute insanity, is a question of another kind, and which we can scarcely hope for unanimity of sentiment upon.

There are, it is true, fleeting attacks of despondency which, for a time, reduce the patient's mind to a condition of melancholia. The stomach, in such cases, is mainly in fault, and the spirits are relieved by a simple dose of medicine.

There are other attacks again, whose early symptoms are so extraordinary, that no discrimination or medical knowledge is needed to decide the question of the person's condition. There is such a thing as insanity of the muscles, of which St. Vitus's dance is a well-known example. In this condition the will has no power to direct the action of the limbs.

The mistakes poor invalids thus afflicted are constantly committing are so odd, that they are provocative of laughter rather than of commiseration. Their legs and arms fly out in all directions; their faces, when they would look grave, put on the oddest grimaces. In short, it seems as though a demon were in possession, distorting every action, and forcing the unhappy individual to make contortions against his wish; for in this disease it is clearly against his will that his body rebels, the mind remaining clear and untouched.

In some cases the physical spasm continues constant in one direction.

Dr. Abercrombie gives the case of a lady who seemed to be imitating the action of a salmon at a salmon-leap. She would occasionally double her body up, and with a convulsive spring, throw herself from the floor to the top of a wardrobe fully five feet high. This fit would only last a certain time. Then she would rotate her head from side to side for weeks together, without showing any sign of fatigue. The propensity to gyrate has in some cases been known to extend to the whole body, the patient rapidly whirling round for a month together continuously. In such cases, there was undoubtedly brain disease of a severe character, implicating most probably the co-ordinating power of the muscles.

A still more extraordinary case is on record, of a girl who was continually attempting to stand on her head, with her legs perpendicularly in the air, continuing this ludicrous action fifteen times in the minute for fifteen hours in the day. The labour incurred by these singular muscular actions would have exhausted half a dozen strong men. yet this delicate girl bore up against them for months, without apparent fatigue.

It is the opinion of many eminent physicians, that the present century has witnessed a very large increase of brain disorders, and that this increase has taken place in an accelerated ratio as the strain upon the commercial and public life of the people has become greater. The intense competition which at present exists among all the liberal professions, the excitement accompanying the large monetary transactions which distinguish the trading of the present day, the gambling nature of many of its operations, and the extreme tension to which all classes of the community are subjected in the unceasing struggle for position and even life, have resulted in a cerebral excitement, under which the finely organized brain but too often gives way.

Dr. Brigham, of Boston, in the United States, gives a most deplorable account of the increase of cerebral disorders in his own country, in which he asserts that insanity and other brain diseases are three times as prevalent as in England. This statement would seem to confirm the notion that go-aheadism —if we may be allowed the term—is straining the mental fabric to its breaking-point. And we must remember that the mischief must not be gauged merely by the number of those who fall by the wayside : there must be an enormous amount of latent mental exhaustion going on, which medicine takes no count of. It is a matter of general observation that the children of men of intellectual eminence often possess feeble, if not diseased brains, for the simple reason that the parents have unduly exercised that organ. What applies to individuals, in a certain modified degree applies to the race. A generation that overtasks its brains is but too likely to be succeeded by a second still more enfeebled in its mental organization, and this exhaustive process must go on increasing if the social causes producing it continue in operation.

We have some means of measuring the magnitude of the evil where absolute lunacy is concerned, inasmuch as we possess official returns to deal with, which gauge its rate of increase or decrease with pretty tolerable accuracy; but we have no such means of ascertaining the nature of the increase of those no less grave disorders of the brain which do not bring the patient under the cognizance of the law. If we could take count of the number of able men who, at the very height of their efficiency, and in the very plenitude of their power, are struck with insidious cerebral disease, such as softening of the brain, and drop out of life as gradually and as noiselessly as the leaf slowly tinges, withers, and then flutters to the ground; if medicine had any system of statistics which could present us with a measure of the amount of paralysis that comes under its observation, or of the apoplectic seizures which so suddenly blot out life;—we should doubtless be astonished at the very large increase which has of late years taken place in affections of the brain. It is just possible that the tendency lately observable in the community to take a little more breath in the race of life, to prolong the annual holiday, and to favour the habit of physical exercise, of which the volunteer movement is a noble example, will do something to check the degenerating process at present undoubtedly going on : meanwhile we must see what we can do to remedy the existing evil. It is, we believe, within the province of art to arrest in its early stages many disorders of the brain, if notice were only given in time; but the golden opportunity is allowed to slip, and disordered function slowly but surely merges into disordered organization. We know full well that at least 80 per cent. of cases of insanity are curable if treated early; and we also know that of those received into the great county asylums scarcely 10 per cent. ever recover. The difference between the two, drop through into the condition of drivelling idiots, or of raving maniacs, simply because the curative influences of medicine have been sought too late. In some of the more obscure and fatal brain diseases, such as cerebral softening, general paralysis, epilepsy, &c., the neglect of early treatment is equally deplorable. The insidious approaches of mischief are often foreshadowed by symptoms so trivial that they pass unobserved by relatives and friends. The person so affected will frequently drop his stick or umbrella in his walk; he will in the slightest possible manner drag one leg, a finger will feel numb, or there will be some slight disorder of the sight.

"In the incipient stages (says Dr. Winslow) of cerebral softenings, as well as in organic disintegrations of the delicate nerve-vesicle observed in what is termed progressive, general, and cerebral paralysis, the patient often exhibits a debility of memory, long before the disease of the brain is suspected, in regard to the most ordinary and most trifling matters connected with the everyday occurrences of life; he forgets his appointments, is oblivious of names of his particular friends, mislays his books, loses his papers, and is unable to maintain in his mental grip, for many consecutive minutes, the name of the month or the day of the week. He sits down to write a letter on some matter of business, and his attention being for a second directed from what he is engaged in, he immediately loses all recollection of his correspondence, and leaves the matter unfinished. In this condition of mind he will be heard constantly inquiring for articles that he had carefully put aside but a few minutes previously."

The memory may be considered one of the most delicate tests of the presence of injury, or the progress of natural decay, in the brain. From the hidden storehouse of impressions which we know to be seated in the cerebrum or greater brain, whilst in a state of vigorous health, by the act of recollection we possess the marvellous power of reproducing the countless tableaux of scenes that have occurred during a long and busy life. Some persons never forget a face they have once seen, others will acquire with extreme rapidity a dozen languages containing hundreds of thousands of words, and store them for immediate use ; the musician catches the floating notes of song, and they remain for a lifetime deeply graven on his memory. The artist packs away within his brain the image of the faintest flush of sunset, or the thousand shades of sky, and reproduces them years after on his easel. It may be imagined that a tablet so sensitive to receive and so strong to retain an incredible number of images in a state of health is not unlikely to speedily make a " sign " of its impaired condition. A flaw in an Egyptian slab covered with hieroglyphics is pretty sure to obliterate some of them, and experience proves that brain injury is speedily shadowed forth by defects, more or less grave, of the memory.

In the whole range of psychological inquiry, there is nothing more remarkable perhaps than the "vagaries," if we may be allowed the term, played by the deteriorating agent in the storehouse of memory : sometimes it enters and for years annihilates the vast collection in an instant, only to restore them again as perfect as before; at other times it obliterates group after group of associated ideas in succession, according to the order in which the brain has acquired them. Again a single letter in a word is all that the destroying power lays its hands upon among the immense magazine at its mercy. A study of mental disease affords us many illustrations of the eccentricities presented to us by impaired and morbid memory; among the most remarkable of which is a case we have before mentioned related by Dr. Graves, of Dublin.

The control of the healthy brain over minutiæ of memory, and the automatic manner in which it is exercised, we are all familiar with, but in disease slips of transposing letters escape notice altogether. The records of psychological medicine are full of instances of defects of memory equally trivial consequent upon lesions of the cerebrum. Thus, an old soldier, after suffering a loss of brain-matter from an operation, was found to have forgotten the numbers five and seven; and a schoolmaster, consequent upon a brain-fever, lost all knowledge of the letter F.

Whilst disease touches the memory in this delicate manner, in its more active phases it seizes the organ with a rude and stifling grasp, and removes at once whole masses of carefully acquired knowledge. An Italian gentleman, master of three languages, struck with the yellow fever, exhibited, in the course of it, remarkable phenomena. At the beginning of his attack he spoke English, the language he had acquired last, in the middle of it French, and on the day before his death his native tongue. The total abolition of an acquired language is not at all an uncommon thing in brain disease, and as a rule the memory in such cases may be said to recede to those ideas engraven upon the memory in childhood.

Those persons who have talked a foreign language all their lives, will be found to pray before death in their native tongue. There have been some remarkable exceptions to this rule, however, and, as we have said before, Dr. Johnson, when dying, is said to have forgotten the Lord's Prayer in English, but to have attempted its repetition in Latin. Possibly the explanation of this exception may be found in the fact, that he thought habitually in Latin. There are not wanting instances, however, to prove that the memory, under disease, oscillates between the past and the present. For instance, Dr. Winslow records a case in which a gentleman, after a serious attack of illness, lost all recollection of recent events—his memory presented the tablet engraven with the images and ideas of his youth only; as he gained strength, however, the old and forgotten ones revived. A still more remarkable instance of loss of memory and its sudden resuscitation we quote from the same author.

"Reverend J. E., a clergyman of rare talent and energy, of sound education, while riding through his mountainous parish, was thrown violently from his carriage, and received a violent concussion of the brain. For several days he remained utterly unconscious; and at length, when restored, his intellect was observed to be in a state like that of a naturally intelligent child, or like that of Casper Hauser after his long sequestration. He now in middle life commenced his English and classical studies under tutors, and was progressing very satisfactorily; when, after several months' successful study, the rich storehouses of his memory were gradually unlocked, so that in a few weeks his mind resumed all its wonted vigour, and its former wealth and polish of culture. . . . The first evidence of the restoration of this gentleman's memory was experienced while attempting the mastery of an abstruse author, an intellectual effort well adapted to test the penetrability of that veil that so long had excluded from the mind the light and riches of its former hard-earned possessions."

It would seem as though ideas were registered on the brain in successive layers, the last lying uppermost; and that as the nervous energy retreated, either as a consequence of disease or of gradual decay, so those ideas lost life *downwards*. The condition of the circulation of the blood through the brain in all probability has much to do with these changes in the vividness of the memory, as it is a known fact that some people recollect better by holding the head downwards.

It is observable again that in morbidly active conditions of the cerebral circulation, such as occur in fever and on the approach of apoplexy, the memory is exalted in an extraordinary manner, and events are remembered with a vividness that is almost painful. In the rapid rush of the blood through the brain, that occurs in some excited stages of insanity, it has been remarked that patients have given signs of faculties which they had never evinced in a state of sanity; prosaic persons have suddenly become poetical, and those who normally had no head for figures, have in these conditions shown no ordinary aptitude for them. It would seem as though the blood, when at this high pressure, had penetrated portions of the brain hitherto but feebly supplied, and brought into cultivation cerebral wastes that were before barren. The practical conclusion to be drawn from these sudden lightings-up in old persons of the memory should excite grave attention, as indicative of approaching fatal apoplexy. Sometimes the memory, not only of the idea upon which the mind was last occupied, but the very action of the muscles arising out of it, has been retained in the mind like a fly in amber. Thus a young girl of six, whilst catching playthings thrown by a companion seated on the pavement, fell and received a cerebral concussion, which rendered her insensible for ten hours. When she opened her eyes she jumped to the head of the bed, and asking "Where did you throw it?" immediately commenced throwing little articles of her dress from the bed, exclaiming, "Catch these!" and from that moment was perfectly restored.

The exactitude with which the fractured ends of the severed idea fit,—severed as we have seen sometimes for years,—is very remarkable, and goes to prove that there must be in such cases an instantaneous arrest of the action of the nerve-vesicles, without morbid change however, otherwise they could not at a moment's notice resume their operation at the exact point at which they left off. We can only liken this extraordinary phenomenon of arrest of mind to some accident which has suddenly stopped a machine—the driving band has perhaps suddenly slipped off—and in this instance the driving band in all probability was the circulation of the blood through the brain—the motive power restored, the machine went on as before.

That mechanical pressure upon the surface of the brain, which means an exercise of control over its circulation, according to the degree in which it is exercised, will produce different mental conditions from perfect coma to perfect sensibility—is well known. A man in Paris once made a living by allowing curious physiologists to make experiments of this nature upon him. He had suffered the operation of trephining, and his brain was covered by a thin membrane only, by applying graduated pressure upon which, the man's relations with the whole external world could be cut off and restored by the mere action of the finger. At the will of the operator he lived alternately the life of the highest order of animal, or that of a mere vegetable.

There is a very remarkable condition of brain, in which the mind of the individual is possessed with a double consciousness. Alternate states arise, as distinct in themselves as though they belonged to two individuals. Dr. Mitchell relates a case of this kind which is so extraordinary that we must be pardoned for quoting it entire :—

" Miss R——, possessing naturally a very good constitution, arrived at adult age without having it impaired by disease. She possessed an excellent capacity, and enjoyed fair opportunities of acquiring knowledge. Besides the domestic arts and social attainments, she had improved her mind by reading and conversation, and was well versed in penmanship. Her memory was capacious, and stored with a copious stock of ideas. Unexpectedly and without forewarning, she fell into a profound sleep, which continued several hours beyond the ordinary time. On waking, she was discovered to have lost every trace of acquired knowledge. Her memory was a *tabula rasa;* all vestiges, both of words and things, were obliterated and gone. It was found necessary for her to learn everything again. She even acquired, by new efforts, the art of spelling, reading, writing, and calculating, and gradually became acquainted with the persons and objects around, like a being for the first time brought into the world. In these exercises she made considerable progress. But after a few months another fit of somnolency invaded her. On rousing from it she found herself restored to the state she was in before the first paroxysm; but she was totally ignorant of every event and occurrence that had befallen her afterwards. The former condition of her existence she called the old state, and the latter the new state; and she was as unconscious of her double character as two distinct persons are of their respective natures. For example, in her old state she possessed all her original knowledge; in her new state only what she acquired since. If a gentleman or lady were introduced to her in the old state, and *vice versâ* (and so of all other matters), to know them satisfactorily she tried to learn them in both states. In the old state she possessed fine powers of penmanship, while in the new state she wrote a poor awkward hand, not having time or means to become expert. During four years and upwards she underwent periodical transitions from one of these states to the other. The alternations were always consequent upon a sound sleep. Both the lady and her family were capable of conducting the affair without embarrassment. By simply knowing whether she was in the old or new state, they regulated the intercourse and governed themselves accordingly."

If there is any truth in our hypothesis of the memory of impressions lying in layers, superimposed one upon another, on the surface of the brain, the alternation of the child-like and the adult state of intelligence would be accounted for by supposing that the level of the power that vivified the nerve vesicles stamped with the mental impression stood at different periods at different heights, retreating in the child-like state to the lowest ebb, and again remounting to its full intellectual height in the adult period.

There is no circumstance with regard to the human economy more remarkable than the tolerance sometimes exhibited by the brain, of grave lesions and disorders within its substance. The popular idea that to touch the sensorium is tantamount to annihilating the life, is a monstrous fallacy. A surgeon lately informed us that he had a young stable-boy lately under his care, whose skull had been fractured by the kick of a horse, which forced it in upon the cerebral mass, and so crushed it, that a portion had to be removed; nevertheless, the patient recovered, and it was remarkable that whereas, before the accident he had been subject to fits, and was rather a dull boy, after the accident he became much brighter, and continues so to this day. In all probability these fits were of an epileptiform character, owing to the pressure of a specula of bone upon the surface of the brain, and when this was removed by the operation, the cause that led to his dulness no longer existed. The kick of the horse was in fact the most fortunate thing that could have happened to him.

Dr. Ferrior relates the case of a man who retained all his faculties entire until the moment of his death, yet one-half of whose brain was on examination discovered to have been destroyed by suppuration. Dr.

Heberden tells us of a man who performed the ordinary duties of life with half a pound of water resting on his brain!

Nevertheless, we are inclined to believe that even in these anomalous cases there must have been some disturbance of the mental powers observable, had the attention of a competent observer been directed to them; and that as a rule it will be found logically true that wherever there has been found the trace of organic cerebral change, there also must have been manifestations of mental disturbance. It is not often that fracturing the skull proves a curative operation, but there can be little doubt that mere accidental shocks to the sick brain have proved far more effective than even the skill of the physician.

We have it on the authority of Petrarch, that a slight concussion of the brain wonderfully strengthened the memory of Pope

Clement VI. It is equally certain that tumours have gone on slowly increasing within the substance of the brain itself without for a long time disturbing the mental power of the individual.

In the great majority of cases, however, *post-mortem* examinations present but faint signs of any lesion of substance, even where the mind during life has been thoroughly disordered. The physician but too often seeks in vain the lunatic's brain for any trace of disorganization. He knows, nevertheless, that alterations of some kind must exist, and attributes his failure to the coarseness of the methods of examination at present employed. The scalpel alone will never find it out, and even the microscope as yet fails to detect departures from normal structure of so delicate a kind as those which are sufficient to overturn noble minds; and we believe that, in order to detect the more subtle lesions of the brain, we must call in the labours of the Chemico-Cerebral pathologist.

Sir B. Brodie has shown that the nervous substance of the brain is distinguished from all other tissues (the bones excepted) by the very large proportion of phosphorus which it contains, amounting to no less than 1.5 per cent.; and if we speak of the solid matter alone, the important position held by this chemical agent in the brain is still more apparent, no less than one-tenth of the whole being composed of phosphorus.

It is a well-known fact, that any laborious mental exercise, indeed any protracted exertion of the nervous system, results in a discharge of large quantities of the phosphatic salts by means of the kidneys. This circumstance, taken together with the remarkable fact that in the brain of the adult idiot there is a very small amount of phosphorus—not more than in that of a child—points to the conclusion that it plays a very important part in the substance of the mental powers.

That in the large majority of the cases of insanity the blood is mainly in fault, there can be little doubt; but when we remember how slight an alteration in the constitution of the vital fluid will produce cerebral symptoms of a very marked character, we no longer wonder at the pertinacity with which these changes have eluded our observation. There are certain moments before dinner when most men suffer what the late Dr. Marshall Hall called the temper disease, the amiable become suddenly unamiable, and the best of us snappish; the *morale* of the individual is entirely altered. Want of rest, again, will so exhaust the mind, that people positively are subject at such times to delusions, imagining their best friends are slighting them, and exhibiting in various ways quasi symptoms of insanity.

We very much question, however, if chemists yet possess skill enough to detect the temporary errors of the blood, which we know must have given rise to this condition of things. Let us ask again, In what particular does the blood differ during sleep from that which it presents in the waking state? It contains, we know, a trifle more carbonic acid; but surely this addition will not account for the act of dreaming, in which we rehearse, as it were, in the inner world of the brain, the wildest thoughts of the insane.

If the pathologist is so often baffled in detecting actual disorganization of the instrument through which mind is manifested, the alienist physician is rarely at a loss to read the symptoms that during life are sure to present themselves. The public are apt to date the amount of mental disturbances from some overt act, which has startled and compelled the attention of friends. Alas! the first overt act, in too many cases, has also been the last, and the verdict of suicide committed in a fit of temporary insanity is considered sufficient to exonerate all parties from any blame; but in every case the first overt act has been preceded by signs and portents of the patient's state of mind, which the experienced eye cannot fail to detect. Only lately the Church had to deplore the suicide of a very able chancellor of a western diocese. On the inquest it was stated that he had been troubled in his mind for several days previous to the catastrophe by an error of 2s. 7d. which he had made in his diocesan accounts. This symptom of a departure from the well-known ordinary masculine tone of his mind would have suggested to any skilful physician the necessity for having him placed under surveillance; had such a step been taken, his friends probably would not have had to lament his loss.

It may be urged, we know, that if we refine too much in this direction, the merest effects of temper and exhibitions of eccentricity, which constitute character, will at last be looked upon and watched with suspicion, as indicating a tendency to mental disease, and that those only will be considered to be sane who possess ordinary level minds, without sufficient originality to go out of the beaten track. Such an error in reasoning no well-educated physician would be guilty of; but he would note with extreme suspicion any sudden change of a man's settled habits or revolution in his modes of thought. As Dr. Andrew Combe remarks :—

"It is the prolonged departure, without any adequate external cause, from the state of feeling and mode of thinking usual to the individual when in health, that is the true feature of disorder in mind; and the degree in which this disorder ought to be held as constituting insanity is a question of another kind, and which we can scarcely hope for unanimity of sentiment upon."

There are very many cases, however, in which insanity shows itself by a simple exaggeration of usually healthy conditions. In these cases the physician finds the greatest difficulty in saying where the line shall be drawn which shall bring the patient under the eye of the law. The naturally passionate man becomes outrageous, the religious becomes fanatical, the vain exceedingly boastful, the liberal extravagant; the only departure from the ordinary mental condition in these cases, is an extraordinary exaltation of the passions and emotions.

It is cases such as these which produce so much misery in the domestic circles, inasmuch as the present state of the lunacy law does not justify their being placed under control. A person thus affected may with impunity squander his whole substance and bring his family to ruin; he may render them miserable for years by the most unfounded suspicions;

he may bring disgrace upon his name by exercising that excess of the secretive power which finds its climax in meaningless petty thefts. The conditions of sanity and insanity in such cases graduate so imperceptibly into each other, that the physician scarcely dares to give a certificate of insanity; and many families are forced to stand idly by whilst they see themselves irretrievably devoted to ruin, merely because the rigid rules of the lunacy law cannot be made flexible enough to meet the ever-varying phenomena of diseased mind.

The difficulty of discovering the physical cause of many forms of insanity is easily accounted for, if there is any truth in the hypothesis that there is such a thing as a co-ordinating mental power, the disease of which is liable to produce the strongest psychological eccentricities. The later psychologists hold that the physical actions are governed, as it were, by a special power which is believed to reside in the cerebellum, or lesser brain; and the disease popularly known as St. Vitus's Dance is supposed, on very good grounds, to arise in consequence of a derangement of that power. The patient cannot conduct the food to his mouth; his legs go every way but the right one when he attempts to walk; he makes the oddest grimaces when asked to look you in the face; and, in short, is so incapable of performing one act of volition as he should do, that the disease is aptly called "the insanity of the muscles."

Having contemplated the frightful effect of disease of the co-ordinating power, let us for a moment consider the exquisite nicety with which that power, when in health, adjusts the muscles to perform any specific act. Let us take, for example, the muscles of the arm of Paganini, in drawing forth the exquisite tones of his violin. It is almost impossible to conceive the precision and *aplomb* with which different groups of muscles must have been directed to produce the delicate shades of music he called forth by a simple act of volition; yet this accuracy, however often repeated, never failed him. Let us grant that there is some co-ordinating power—some executive presiding over the just association of our ideas—and there is no incoherence for which its disease may not be held responsible.

"There is no fixed or even transient delusion," says Dr. Winslow, in the case of physical chorea. "In these cases the insanity appears to depend upon a disordered state of the co-ordinating power (eliminated, in all probability, in the cerebrum), and paralysis of what may be designated the executive, or, to adopt the phraseology of Sir William Hamilton, regulative, or legislative faculties of the mind. The patient so affected deals in the most inexplicable absurd combinations of ideas. Filthy ejaculations, terrible oaths, blasphemous expressions, wild denunciations of hatred, revenge, and contempt, allusions the most obscene, are often singularly mingled with the most exalted sentiments of love, affection, virtue, purity, and religion . . . I have often known patients whilst suffering from the choreic type of insanity alternately to spit, bite, caress, kiss, vilify, and praise those near them, and to utter one moment sentiments that would do honour to the most orthodox divines, and immediately afterwards to use language only expected to proceed from the mouths of the most depraved of human beings. This phase of mental aberration is often seen unassociated with any form of delusion, hallucination, or illusion."

What the nature of this mental regulative force may be we know no more than we do of the muscular co-ordinating power. Physical methods of inquiry tell us nothing, and cannot be expected to do so.

It has been said by Cicero, that if it had been so ordered by nature, that we should do in sleep all we dream of doing, every man would have to be bound down before going to bed. It does seem remarkable that during one-third of our lives we should be liable to a derangement of the mental power (for such is dreaming), which in our waking state would render us liable to be placed in a lunatic asylum.

The very intimate connection undoubtedly existing between dreaming and insanity has in all times attracted the attention of psychologists, and, of late, physiologists have directed their attention to the physical conditions which give rise to the former very remarkable state.

Dr. Marshall Hall believed that sleep is produced either by some constriction of the great vessels of the neck, or by a sluggishness of the respiratory organs, either cause leading to a venous condition of the blood calculated to produce somnolency. We know that every degree of insensibility, up to complete coma, can be produced by simply allowing the neck to rest with the weight of the trunk against a tightened cord. Nature has, therefore, only to contract the great vessels periodically, to bring about the state of things we so readily do artificially; but sleeping is not dreaming, says the reader. Certainly not; but it is the dark background on which the pattern of our dreams is woven, and in all probability the condition of the circulation through the brain which produces it, is also answerable for the diversified pattern itself.

The absence of volition, says Dr. Darwin, distinguishes the state of sleep from the waking state. This proposition is, however, rather too sweeping; for in all probability there is no such thing as perfect sleep, or absence of volition, any more than there is any position in which every muscle of the body is totally at rest; at all events, in dreaming there are many reasons which lead us to conclude that the different portions of the brain sleep unequally, and this inequality possibly arises from the position of the head, directing a fuller flow of blood to one part of brain than to others, or from its detention in given portions. If we examine a dream narrowly, we find that volition may or may not be excited, according to the nature of the excitement created in the mind by the illusion passing before it.

For instance, it often happens that we dream we are pursued by a mad bull or by an assassin, and the greatest distress is occasioned by finding that we can neither call out nor run away. It again often happens to us that we dream we are suddenly falling down a precipice; but here volition is, as it were, suddenly awakened out of its sleep, for we find that in the endeavour to save ourselves from falling, we jump *up* in the bed.

We have here a proof that volition does not rest so soundly, but that it can be roughly and suddenly shaken into life. In somnambulism it is actively awake, although consciousness is perfectly dormant. There is also such a thing as day-mare—a condition of the brain which exists just as we are waking from sleep, when we are perfectly conscious, but unable either to move or to call out; volition in fact has slept longer than the other faculties of the brain.

It is noteworthy, that sleeping on the back is generally assigned as a cause of nightmare, or that condition in which action seems most obstinately bent upon not answering the appeals made to it. This fact certainly seems favourable to a belief that position has something to do with the unequal manner in which the different faculties of the brain rest during sleep. The seat of the muscular co-ordinating power, the cerebellum, in the recumbent position may possibly suffer congestion in consequence of its lying partially under the cerebrum. The state of reverie or of daydreaming presents many features which are very analogous to that of mental aberration. Except that we are conscious of our abandoning the fancy to its own will, this condition differs but little from that of dreaming.

An indulgence in this habit tends to emasculate the mind. When long continued it is often precursory of softening of the brain, and of the incipient stages of some types of mental disorders. Disraeli, in his "Contarini Fleming," has with intuitive genius seen this truth :—

> "I have sometime," he says, "half believed, although the suspicion is mortifying, that there is only a step between his state who deeply indulges in imaginative meditation, and insanity; for I well remember when I indulged in meditation to an extreme degree, that my senses appeared sometimes to be wandering. I cannot describe the peculiar feelings I then experienced . . . . but I think it was that I was not always assured of my identity, or even existence; for I found it necessary to shout aloud to be sure that I lived; and I was in the habit very often at night of taking down a volume and looking into it for my name, to be convinced that I had not been dreaming of myself."

We may allude to one faculty of the brain which appears always to remain dormant during dreams : we allude to the faculty of wonder. The most incongruous images, the oddest combination of circumstances, the strangest persons, present themselves before us at such times unchallenged. We converse with friends and relations long since dead, without feeling the least surprised at their resurrection. And why is this? Because the sense of the fitness of things is also wanting. How can we wonder when the standard of judgment is absent? And herein we find the extraordinary likeness between dreaming and certain forms of insanity. The co-ordinating psychical power in both cases is in abeyance. Sir Walter Scott has shrewdly said, that the only difference between the two states is, that in dreams the horses have run away with the coach whilst the coachman is asleep; in lunacy the runaway takes place whilst the coachman is drunk. This distinction is a nice one, but the effect upon the coach in the two cases is so remarkably alike, with the single exception of the absence of volition in the former, that we think the psychologist is justified in considering them associated phenomena of mind.

There have not been wanting cases, indeed, in which the first outbreak of insanity commenced in a dream.

" A gentleman (says Dr. Winslow) who had previously manifested no appreciable symptoms of mental disorder or even of disturbed and anxious thought, retired to bed apparently in a sane state of mind; upon rising in the morning, to the intense horror of his wife, he was found to have lost his senses ! He exhibited his insanity by asserting that he was going to be tried for an offence which he could not clearly define, and of the nature of which he had no right conception. He declared that the officers of justice were in hot pursuit of him ; in fact, he maintained that they were actually in the house. He begged and implored his wife to protect him. He walked about the bedroom in a state of great apprehension and alarm, stamping his feet and wringing his hands in the wildest agony of despair. Upon inquiring into the history of the case, his wife said that she had not observed any symptom that excited her suspicion as to the state of her husband's mind; but upon being questioned very closely, she admitted that during the previous night he appeared to have been under the influence of what she considered to be the nightmare or a frightful dream.

Whilst apparently asleep, he cried out several times, evidently in great distress of mind—'Don't come near me! Take them away! Oh save me, they are pursuing me!' It is singular that in this case the insanity which was clearly manifested in the morning appeared like *a continuation of the same character and train of perturbed thought that existed during his troubled sleep*, when, according to his wife's account, he was evidently dreaming."

Sir Benjamin Brodie has referred, in his Psychological Inquiry, to a very remarkable quality in the brain, a quality Dr. Carpenter terms unconscious cerebration. It often happens that, after accumulating a number of facts in an inquiry, the mind becomes so confused in contemplating them, that it is incapable of proceeding with its labours of arrangement and elaboration; dismayed at the chaotic heap, it backs as it were upon itself, and we feel certain that it is of no use cudgelling our dull brains any longer. After a little while, however, without having once consciously recurred to the subject, we find to our surprise that the confusion which involved the question has entirely subsided, and every fact has fallen into its right place.

Is it possible that the brain can, without our knowledge, select and eliminate, aggregate and segregate facts as subtilely as the digestive organs act upon the food introduced to the stomach? Sir Henry Holland is inclined to dissent from such a conclusion, and leans rather to the explanation of the phenomenon which Sir B. Brodie has himself suggested; viz., that the seeming ordering process may be accounted for by supposing that all the unnecessary facts fade from the memory, whilst those which are essential for the ultimate arrangement and classification of the subject under consideration are left clear of the weeds that before encumbered them. But does not this explanation involve a confession of an eliminative process going on unconsciously in the brain, which appears to be little less wonderful than a hidden cogitation? Why should the unessential facts alone fade? Why should we refuse to recognize masked operations of mind? Surely we see every day examples of cerebral acts being performed of which the individual is afterwards totally oblivious. Let us instance, for example, the mental impressions engraved with a searing-iron, as it were, upon the brain in the moments of delirium.

Under chloroform, again, the mind is often in a state of great exaltation, and goes through mental labour of a kind calculated, one would imagine, to leave lasting traces behind it on the memory; nevertheless water does not more readily give up impressions made upon it than does the tablet of the brain under its influence. Even in dreams, of which we take no note, but which are patent to bystanders by our speech and actions, there must be plenty of "unconscious cerebration." Indeed, Sir Henry Holland, in referring to a vague feeling that all of us at times have experienced when engaged in any particular act, that "we have gone through it all before," endeavours to explain it by supposing that the faint shadow of a dream has suddenly and for the first time come to our recollection in a form so unusual that it seems as though we had acted the part before in another world. That we go through brain-work unconsciously we have therefore no doubt; and we see no reason why we should deny the existence of a power seated in the brain, whose duty it is silently to sift the grain from the husk in the immense mass of mental pabulum supplied to it by the senses.

There can be found no more curious chapter in the history of the human body and mind than that which relates to the phenomenon of morbid attention directed to its different organs. The power of influencing any particular portion of the animal economy by the concentration of our attention upon it, is so marvellous, that we wonder the method of its action has not been more thoroughly investigated than it appears to have been. It would seem as though the mind possessed the power of modifying the functions of distant parts of the body, and of exciting sensations quite independently of any act of volition.

The mere act of attention to any particular organ over which we possess no muscular control, is sufficient to produce some alteration of its functions. Thus we may will that a spot in the skin shall itch, and it will itch, if we can only localize our attention upon the point sufficiently; by directing our thoughts to the heart, it rapidly beats; by soliciting the lower intestine, it is quickly brought into action. There is scarcely an organ of the body which is not liable to be interfered with by simply concentrating the attention upon it. Whole regions of superficial nerves, such as those of the skin in the neck, may be exalted in their action to the highest degree at the mere expectation of being tickled there.

This nervous attention may become so persistent as to cause actual disease. We have a familiar instance in dyspepsia, where the patient is for ever thinking of his stomach, and at last diseased function degenerates into diseased organization, and he falls into the condition of a helpless hypochondriac. But if an attitude of concentrated attention upon his mere animal functions is thus capable of producing disease in them, what effect has it upon the mind itself? Sir Henry Holland has very subtly remarked, that it appears to be a condition of our wonderful existence, that while we can safely use our faculties in exploring every part of outward nature, we cannot sustain those powers when directed inward to the source and centre of their operations — in other words, the mind, when it persists for any length of time in analyzing itself, scorpion-like, stings and destroys its own action.

That we can as readily injure our brains as our stomachs by pertinaciously directing our attention to fancied diseases in them, cannot be doubted; and that mere perversion of ordinary modes of thought, such as may exist in minds only functionally disordered, may be fixed by the action of morbid attention, so as to constitute permanent aberration, is equally certain.

And let it not be supposed that it is a rare thing to find reason struggling manfully with the promptings of insanity. Bishop Butler tells us that he was all his life struggling against devilish suggestions, and nothing but the sternest watchfulness enabled him to beat down thoughts that otherwise would have maddened him. His case was but an example of that of thousands of persons with whom we come in contact every day, who, under a calm exterior, conceal conflicts between the reason and the first promptings of insanity of the most terrible kind,—persons, in fact, who are hovering on the borderland of insanity, and may at any moment pass it.

HALLUCINATIONS AND DREAMS.

———0———

THE hallucinations of all ages and countries have been marked by one invariable fact; they have referred to some particular train of thought or religious sentiment that impressed the public mind, and the age in which such visions have been seen. Thus, in the old Roman time, the apparitions witnessed, or supposed to have been witnessed, referred to the pagan deities, the fairies and satyrs, and other ideal beings by which the ancients personified nature in her different attributes. In the Christian era, the visions referred to the Almighty, Christ, or the Devil, or those sacred personages that were made the subject of their daily thoughts.

The Arabians, again, had visions of genii and gins, and the different nationalities kept to those reflections, if we may so speak of the current ideas of the day. The Christian never saw Bramah, and the Brahmin never was allowed to witness the apparition of Christ or the Virgin Mary; neither did the Catholic ever see genii or a satyr. They saw only those personages who were in their daily thoughts, or who were associated with the deepest emotions of their nature. This is, we think, the proof that each vision proceeded from the minds of the spectators, instead of having been really seen from without. Dr. Bouirre de Bisment, in his work on 'Hallucinations,' has collected a vast number of cases illustrative of this position, and of the various mental conditions under which persons have seen visions and spectres. It is said of the great Talma that it lay within his power to transform his brilliantly dressed audience into so many skeletons. Dr. Wigan says, that an artist, who succeeded to a large portion of the practice of Sir Joshua Reynolds, turned the faculty of seeing his sitters after they had left to great pecuniary advantage. He was in the habit of taking only half an hour's sitting, during which time he studied the individual minutely, and in this way so impressed his features upon his memory that he never required his presence again. When he wished to proceed with the portrait, he called up the sitter, and placed him in a chair, and so finished the portrait in a very short time. Several sitters were at the same time stored up in his memory, and were so called up at his will. This extraordinary power was, however, a symptom of coming disease; for after he had taken advantage of it for several years, during which time he never painted less than three hundred portraits a year, his faculty began to fail him, his imaginary sitters began to dispute respecting the real sitters to which they belonged; he became confused, and ultimately insane. He described his power of seeing the sitter after he was gone as more vivid than the real life; but all these exaltations of the senses are premonitary of disease. Dr. Wigan gives us the experience of another patient, who had the power of placing himself before his own eyes. This double laughed when he laughed, and even argued with him. Thus, haunted by himself, at last it grew beyond a joke; he became miserable, and resolved to terminate his existence. He did this with great deliberation; for he waited until the end of the year, when, upon the night of the 31st December, having made up all his accounts, he shot himself. Now, although in neither of these cases was there any insanity in the outset of these singular examples of hallucination, yet the termination of them clearly proved that the minds of these individuals were not in a healthy state; they were, in fact, suffering from the incipient stage of brain disease, which is wholly undiscoverable to friends and casual acquaintances.

Hallucinations involving insanity are often described by those experiencing them in a manner so circumstantial that, were it not for the absurdity of their statements, it would be difficult to doubt that they were made truthfully and in good faith, and had really foundation in fact.

Dr. Prichard gives an example from the mouth of a patient, which is so life-like that it seems a pity to omit one word of it. He said :—" One afternoon, in the month of May, feeling himself a little unsettled, and not inclined to business, he thought he would take a walk into the City to amuse his mind; and having strolled into St. Paul's Churchyard, he stopped at the shop-window of Carrington and Bowles, and looked at the pictures, among which was one of the Cathedral. Here he met with an elderly gentleman, dressed in dark brown clothes, who entered into conversation with him, and persuaded him to dine with him, and afterwards to ascend the ball of St. Paul's just below the cross.

"They had not been there many minutes when, while he was gazing on the extensive prospect and delighted with the splendid view below them, the grave old gentleman pulled out from an inside coat-pocket something like a compass, having round the edge some curious figures; then, having muttered some unintelligible words, he placed it in the centre of the ball. He felt a great trembling, and a sort of horror came over him, which was increased by his companion asking him if he should like to see any friend at a distance, and to know what he was at that time doing; for if so, the friend of the latter should show him any such person. It happened that his father had been for a long time in bad health, and for some weeks past he had not visited him. A sudden thought came into his mind, so powerful that it overcame his terror, that he should like to see his father. He had no sooner expressed the wish, than the exact person of his father was immediately presented to his sight in the mirror, reclining in his arm-chair and taking his afternoon sleep. Not having fully believed in the power of the stranger to make good his offer, he became overwhelmed with terror at the clearness and truth of the vision presented to him, and entreated his mysterious companion that they might immediately descend, as he felt himself very ill. The request was complied with, and, on parting under the portico of the northern entrance, the stranger said to him 'Remember, you are the slave of the Man of the Mirror.'

"I inquired in what way the power was exercised? He cast on me a look of suspicion mingled with confidence, took my arm, and, after leading me through two or three rooms and then into the garden, exclaimed, 'It is of no use; there is no concealment from him, for all places are alike open to him. He sees us and hears us now.' I asked him where the man was that heard us? He replied, in a voice of deep agitation, 'Have I not told you that he lives in the ball below the cross on the top of St. Paul's, and that he only comes down to take a walk in the churchyard, and get his dinner in the house in the dark alley.' He also spoke of the tyranny he exercised over all those within the circle of his hieroglyphics. I asked him what these hieroglyphics were, and how he perceived them? He replied, 'signs and symbols, which you, in your ignorance of their true meaning, have taken for letters and words, and reading, as you have thought, *Day and Martin* and *Warren's Blacking!* Oh, that is all nonsense! They are only the mysterious characters which he traces to mark the boundary of his dominions, and by which he prevents all escape from his tyrannous power. How have I toiled and laboured to get beyond the limits of his influence! Once I walked for three days and nights, until I fell down under a wall exhausted by fatigue and want of sleep; but awaking, I saw the dreadful signs before my eyes, and I felt myself as completely under his infernal spells at the end as at the beginning of my journey.'"

This is the complete hallucination of a lunatic; a story of the most circumstantial nature is concocted, which is yet of the most absurd nature, having no reference to the daily habits of the person, who believes he has been subject to it. In some individuals the visions that appear are of the most simple kind, being generally confined to one person or thing, the imaginative faculty being entirely wanting in them. In the hallucinations that occur in those suffering from *delirium tremens*, as a rule, the visions are in the form of animals running about the room or over the bed, making grimaces; sometimes they appear in extraordinary numbers, but so small that a host will appear upon a sheet of paper. Persons will appear to follow patients affected in this way, who immediately disappear when any attempt is made to clutch them. Any chance object seems to give rise to hallucinations in this disease. The person suffering from these delusions may be perfectly rational on every other matter. He may converse with calmness, and very rationally; but in the midst of his conversation, some portion of the dress of the person with whom he has been conversing, suddenly takes the form of some creeping thing, or of a grinning devil, who flies away with horror. These hallucinations are of a very fleeting character—a fresh potation will often banish for the time the phantoms that appear to surround him.

The effects of opium-eating give rise to the same kind of hallucinations. The visions that are called up by the excessive use of this drug are of a more pleasurable kind. Indeed, De Quincy, in his "Confessions of an Opium-Eater," has given us a picture of the delights experienced by indulging in this narcotic, which would be tempting enough to some minds, did not the miserable condition of the patient in his waking state serve as a horrible warning. Dr. Porqueville, in his "Voyage en Morée," gives a description of a case related by an English

ambassador of an Indian king. This high
personage having been led to a sumptuous
apartment, in a short time two servants bearing a litter approached. Upon the litter,
covered with a shawl of great value, a human form was borne, to all appearance dead.
Presently, however, an officer in attendance
produced a bottle in which a bluish-looking
liquor appeared. The ambassador, thinking
he was the involuntary witness of some
funeral ceremony, wished to retire; but he
was undeceived upon observing one of the
officers raise the head of this apparently life-
less being, replace the tongue, which was
hanging from the mouth, and make it swal-
low some black liquid, at the same time
closing the mouth, and gently rubbing the
throat in order to facilitate its passage.
When this operation had been repeated five
or six times, the figure opened its eyes, and
closed its mouth of its own accord; it then

swallowed without assistance another large dose of the liquid, and in less than an hour became revived, and sat up upon the couch, having somewhat recovered its natural colour and the partial use of its limbs. He then addressed the envoy in person, and asked him the object of his mission. For nearly two hours this extraordinary being remained perfectly conscious and capable of transacting business of the greatest importance. The English ambassador took the liberty of asking him some questions concerning the strange scene he had witnessed. "Sir," he replied, "I have long been an opium-eater, and by degrees have fallen into this deplorable condition. I pass three parts of the day in the torpid condition in which you have seen me. Although incapable of moving or speaking, I retain my consciousness, and during this time I am surrounded with the most delightful visions; but I should never awake if I were not surrounded by zealous and affectionate attendants, who watch over me with the most anxious solicitude."

The use of naschisch in the East produces the most delightful visions. It is a preparation of Indian hemp, a very powerful narcotic, and one which is coming into use in this country. The fact, that by the use of drugs we can artificially produce hallucinations for a very short period, such as are persistent in the really insane, is very curious, and proves that in the latter case there is an exaltation of the brain, the product of a morbid condition, produced probably by the blood. It must be confessed, however, that the visions and delusions produced by the use of drugs are different in kind to the true hallucinations of the insane. The mind sees with the inner eye as it were, and the figures or visions partake more of

Hallucinations and Dreams. 271

the nature of those which appear in a dream. When the person under them has recovered from their effects, he is conscious that what he has seen was the product of his own excited brain. No doubt, however, when the habit of taking any drug which acts in this manner upon the brain is persisted in, the visions appear to move in the outward world, just as they do in *delirium tremens*, which is produced by the action of spirits. In all such cases of brain excitement the senses, for the time, are preternaturally acute, the hearing and the sight are marvellously exalted, and the memory for past events and scenes is very vivid.

The scenes that pass before the mind in sleep may be likened to those produced by narcotism. Whilst they are passing like a panorama, they seem to be veritable objects, and we believe in them most implicitly. The most extraordinary events occur without in the least appearing strange to us, indeed, the senses of surprise and comparison seem for the time suspended; judgment is also wanting; in short, we seem to be quite as satisfied of the naturalness and truth of the most extraordinary and contradictory scenes and actions, as do the insane with respect to their own ideas. Indeed, the waking dreams of the demented are in many respects the counterparts of those which we experience in our healthy slumbers; the only difference is that the sane do not act upon them. But in the case of the somnambulist, there is not even this point of difference. Persons in this condition are continually walking in their sleep—generally without harm to themselves, but sometimes they walk through open windows and are killed. A case is told of a monk in a religious house in Germany, who used to walk about the monastery at night in one of these fits. On one occasion he knocked at the Friar's door, who not happening to be in bed, let him in. He immediately made his way to the bed, and, with a knife which he had, stabbed the clothes through to the matress, and then returned to his own cell. The next day, upon the Friar inquiring of him what was passing in his mind when he performed this terrifying act, the monk, much disturbed, replied, " My father, I had so strange a dream that I am most reluctant to tell it to you. It was, perhaps, the work of the devil, and——" " I command you," said the Friar. " My father," he then said, " scarcely had I gone to bed, when I dreamed that you had killed my mother; that her bleeding phantom appeared to me and demanded vengeance. At this spectacle I felt in such a transport of fury that I ran like a madman to your apartment, and having found you I stabbed you. After that I awoke in a profuse perspiration, horrified at my attempt, and I returned thanks to God that I was free from so great a crime." "You were nearer committing it than you imagine," said the Friar, and very discreetly ordered that in future he should be locked in his cell for the night. This unconsciously acted drama has no doubt often been repeated with a more tragic termination in other times, and possibly many a man has suffered the punishment of death for a homicide perpetrated entirely in a state of unconsciousness. It is a remarkable characteristic of the state of dreaming, that the mind often assimilates, in the train of ideas it is pursuing, any chance sound that may strike upon the ear. The slamming of the door for instance, is changed into the discharge of a gun, and the current is no doubt often changed by these interruptions. Thus, it is obvious that the brain has a certain power of directing its action, even when we are totally unconscious. The bodily movements, again, which take place in sleep, set the mind upon a new course of adventure; the excitement which takes place in the different organs suddenly colours the misty action of the dream, and no doubt the extraneous sights and sounds are accountable for many of the sudden distractions which we all experience in the visions we have in the night. Many persons who sleep in snatches have the power of continuing the thread of their dream after it has been broken by the waking state. We have this power ourselves, and we feel that by a constant practice the habit would become so persistent, that it may be transmitted into an act of will, as in the mind's action in the day time.

In the act of dreaming many persons talk, holding imaginary conversations with individuals. Maniacs very often unconsciously give a clue to the cause of their afflictions in this manner; secrets that they have kept close during the day with the reticent cunning of their class, thus leak out in the silence of the night. Esquirol utilized this fact in his asylum by passing the night near the beds of patients whose history was unknown to him, and in this manner possessed himself of a key to their malady. That ideas occur to us in sleep which we are not capable of in our waking state, the experience of every one proves. It is acknowledged that there is such a thing as unconscious cerebration; in other words, when we have cudgelled our brains in vain over some mental work, and are compelled to give it up in disgust, it often occurs that on reverting to the subject next morning our ideas flow from our pen smoothly and swimmingly. This fact will account for the remarkable tales we hear of mental difficulties solved during sleep. It is related of Tartini, the famous composer, that after wearying himself in vainly attempting to finish a sonata, he fell asleep, and dreamed of the theme that was in his mind. In this dream the devil appeared to him and proposed to help him in his sonata, provided he would give him his soul in return. He agreed, and the devil at once composed the sonata off-hand, in the most charming manner. When he awoke he rushed to his desk and put down the notes which still lingered in his memory, and the result was the masterly sonata which is now known as the " Sonata du Diable."

The absence of volition, which as a rule

characterizes the state of dreaming, marks the difference between it and insanity, otherwise the two conditions are wonderfully alike. This was appreciated by the ancients. Cicero says that if it had been ordered by Nature that we should actually do in sleep all we dream of doing, every man would have to be bound down before going to bed. " Half our days we pass in the shadow of the earth, and the brother of death extracteth a third part of our lives," says Sir Thomas Browne.

Some of the hallucinations of the early and middle ages were of the most remarkable kind, and prove that the transmigration of the human body into animals was a popular belief, and was, as far as possible, actually put in practice. Thus Lycanthropy possessed large numbers of people in France and Germany in the fourteenth and fifteenth centuries. Persons suffering from this form

Hallucinations and Dreams. 279

of hallucination imagined that they had married female wolves. They actually left their homes and went into the forests, where they let their hair and nails grow, and became for the time wild beasts. They were known popularly as were-wolves, and it was supposed that they were thus transformed by witchcraft. Persons so possessed became furious beyond the most savage animals; they mutilated and even devoured children. At the trial which took place in 1521, at Besançon, three of these were-wolves confessed that they had given themselves over to the devil. One of these poor creatures said that he had killed a boy with his teeth and claws, and would have eaten him but for fear of the country people; another admitted that he had killed a young girl as she was gathering peas in a garden; another confessed that he had not only killed, but eaten, four others. These poor creatures were evidently insane; but the science of mind in those days could not be expected to have realized this fact, when we find that even now the Law slays when it should put the homicide in an asylum. These three madmen were accordingly burned alive, the fate of all such demented beings who were afflicted with the like delusion in those days. The middle ages, possessed with a belief in the apparitions of angels and saints and devils, were necessarily imaginative ages, and immensely dramatic in their poetic forms. We have not room to afford instances of the countless apparitions which were called forth by the religious spirit of the times. A vast number of these visions or apparitions seen by the hallucinated had no bearing with reference to the movements of the period; most of the warnings they gave were never fulfilled, and were forgotten; but the few which really made a hit were remembered, and handed down from father to son by the cunning spirit of priestcraft: just as in the present day all the minor hits of Zadkiel, in his Almanack, are duly trumpeted forth in the next year's issue, whilst the unfulfilled prophecies, like those of Dr. Cumming, are quietly ignored or adjourned for the benefit of some succeeding generation.

But the power to see visions and to receive spiritual communications from angels is far from having yet departed from among us. The upper classes have, it is true, left off these old romances of the past, but they have taken to other forms of " spiritualism," as we shall presently show. The lower classes still believe in the old story, and repeat with increasing repetition the old delusions inculcated in the middle ages by the priesthood. The latest instance in which the lower middle class has shown a revival of the old spiritual form of hallucination, the "Shakers" may be mentioned, a company of whom has lately been imported from the United States. It seems strange that in the New World, where it would be supposed old ideas would have failed to make a footing, this society should not only exist, but should number its thousands. It may be that the power of imagination was the principal agent in enabling Macdonald, a disciple of Robert Owen, to induce a body of emigrants, many years ago, to go over to that country and to establish a community, living in common after the ideas of their master,—a community who can be brought to believe that such a system is possible, even in America, may easily be imagined to be open to accept any form of spiritual delusion.

The Shakers, it appears, were founded by a factory girl, born in the latter part of the last century in Manchester. Her name was Anne Lee. She was believed to be an incarnation of the Almighty and the mystical Bride of the Lamb referred to in the Revelations. She took to preaching in the streets, caused an obstruction, was locked up in the Bridewell, and during the night she affirmed that the Lord appeared to her, and was spiritually united to her. When she was released, a number of followers surrounded her, and an angel directed them to depart with her to the land of promise, America. They took up their abode in the settlement of Water Violet, in the State of New York. This girl they held as their director, and termed her Mother Anne. After ruling them about eight years she died, and her disciples profess to hear from her constantly from her heavenly abode. These disciples are strictly celibates, and, strange to say, no scandal has ever been reported respecting them. We know what has been reported of the monks of old; and we can only account for this exemption from like reports by supposing that the religious fervour of the new people held them above a physical weakness which is certain in any societies to appear as soon as familiarity overrides the strong original inspiration which induced them to exclude themselves from the world.

As an example of the curious mental condition of these people, we may quote from Mr. Macdonald the peculiar nature of their service, as witnessed by himself.

"At half-past seven, P.M., on the dancing days, all the members retired to their separate rooms, where they sat in solemn silence, just gazing at the stove, until the silver tones of a small tea-bell gave signal for them to assemble in the large hall. Thither they proceeded in perfect order and solemn silence. Each had on their dancing-shoes, and on entering the hall they walked on tip-toe, and took up their positions as follows :—The brothers formed a rank on the right, and the sisters on the left, facing each other, about five feet apart. After all were in their proper places, the chief elder stepped into the centre of the space, and gave an exhortation for about five minutes, concluding with an invitation to them all to go forth—old men, young men, and maidens —to worship with all their might in the dance. Accordingly, they 'went forth,' the men stripping off their coats and remaining in their shirt-sleeves. First, they formed a procession, and marched round the room at double-quick time, while four brothers and sisters stood in the centre singing to them. After marching in this manner until they got a little warm, they commenced dancing, and continued until they were all pretty well tired. During the dance the sisters kept on one side, and the brothers on the other, and not a word was spoken by any of them. After they appeared to have had enough of this exercise, the elder gave the signal to stop, when immediately each one took his or her place in an oblong circle formed around the room, and all waited to see if any one had received a 'gift,'— that is, an inspiration to do something odd. Then two sisters commenced whirling round like a top, with their eyes shut, and continued their motion for about fifteen minutes, when they suddenly stopped, and resumed their places as steadily as if they had never stirred. During the whirl, the members stood round like statues, looking on in solemn silence.

" This whirling process is supposed to screw out of the Mother the message she desires to give to her votaries; and on this occasion the dead Mother Anne was reported by the whirlers to have communicated to them that she had sent two angels to inform them that a tribe of Indians had been around them for a couple of days, and wanted the brothers and sisters to take them in. They were outside the building, looking in at the windows.

"I shall never forget [says Macdonald] how I looked round the windows, expecting to see the yellow faces when this announcement was made; but I believe some of the old folk bit their lips and smiled. It caused no alarm to the rest; but the first elder exhorted his brethren to take in the poor spirits, and assist them to get salvation. He afterwards repeated more of what the angels had said,—viz., that the Indians were a savage tribe, who had all died before Columbus discovered America, and had been wandering ever since. Mother Anne wanted them to be received into the meeting to-morrow night.

"Accordingly, the next evening, after the dancing was over, the two sisters originally intrusted with the message, after the doors and windows of the hall were opened, said that they saw strangers mingling with the brothers and sisters. The chief elder exhorted them that they should take the strangers in.

"Whereupon [says Macdonald] eight or nine of the sisters became possessed of the spirit of the Indian 'squaws,' and about six of the brethren became Indians. Then ensued a regular 'pow wow,' with whooping and yelling, and strange antics, such as would require a Dickens to relate. The sisters and brothers squatted down on the floor together, Indian fashion, and the elders and elderesses endeavoured to keep them asunder, telling the men they must be separated from the squaws, and otherwise instructing them in the rules of Shakerism. Some of the Indians then wanted some 'succotosh,' which was soon brought them from the kitchen in wooden dishes, and placed on the floor, when they commenced eating with their fingers. These performances continued till about ten o'clock; then the chief elder asked the Indians to go away, telling them they should find some one waiting to conduct them to the Shakers in the heavenly world. At this announcement the possessed men and women became themselves again, and all retired to rest."

It is needless to proceed further with this miserable farce; but it is evidence that there is no folly which those possessed with a religious craze cannot be induced to enact. As we have seen, a sect of these Shakers were until lately settled in the New Forest, where a Mother Girling is the leader; and their fanaticism has induced them to suffer severe hardships, in the belief that the Lord has sent them for their good. However benighted these poor people may be, there is something really grand in their behaviour, self-denying as it is, compared with the miserable delusions educated members of the upper classes profess to believe in, to wit, mesmeric influences, table-turning, and spirit-rapping. There can be no possible excuse for such people; and when we see the attempt that was made the other day to declare Mother Girling was insane, we ask ourselves how much more open the drivelling idiots who participate in such miserable delusions are to the charge, than these poor uneducated fanatics, who at least practice the virtue of self-denial.

THE SUICIDAL ACT.

THE notion is firmly fixed in the mind of foreigners that England is the land of suicides. Like many other notions, it is wholly without foundation. Possibly Montesquieu is answerable for this ridiculous idea, as he says, "The English frequently destroy themselves without any apparent cause to determine them to such an act, and even in the midst of prosperity. Among the Romans, suicide was the effect of education; it depended upon their customs and manners of thinking: with the English it is the effect of disease, and depending upon the physical condition of the system."

Now it is not astonishing that even a French philosopher should be ignorant of

English manners and customs, but it is unfortunate that he should have touched upon this topic of suicide to illustrate them, inasmuch as in comparison with the French we are far from being a suicidal nation. We can neither show anything like the number of persons that commit self-destruction, compared with our population, that France can, nor is this awful act committed with us on such slight pretences as it is across the water. Englishmen are supposed to blow their brains out in a fit of the spleen. Foggy weather is said to be a sufficient occasion to create a crop of suicides, and November is believed to be the month which is especially favourable for the commission of this crime. This idea seems to be prevalent even in England, but the statistics of the subject prove that it is directly contrary to the fact. Fogs and damp may depress our spirits, but they do not precipitate us from the Monument or from the bridges. Holland is a much damper, foggier climate than our own, but the Dutch are anything but inclined to suicide.

Paris is the head-quarters of self-destruction. There it would appear to have become an institution. The Morgue is one of the sights of the city. In London the discovery of a dead body in the Thames creates a sensation; in Paris it is a usual thing to see upon the cold slabs of stone three or four dead bodies that have been rescued from the Seine. It may be that drowning is the more common form of committing self-destruction with our neighbours, and we may say there is such a thing as fashion even in this hideous subject.

In this way only can we account for the vast preponderance of French suicides over our own, judging from the experience gathered at the Morgue. The late Dr.

Winslow said " The physician should constantly bear in mind this important fact connected with the suicidal disposition, viz., that those determined upon self-destruction often resolve to kill themselves in a particular manner; and however anxious they may be to quit life, they have been known to wait for months and years until they have had an opportunity of effecting their purpose according to their own preconceived notions. A man who has attempted to drown himself will not readily be induced to cut his throat, and *vice versâ*. A morbid idea is frequently associated in the maniac's mind with a particular kind of death, and if he be removed from all objects likely to awaken this notion, the inclination to suicide may be removed;" and this opinion is endorsed by another eminent authority in lunacy, Sir W. Ellis. Of course there are some cases in which the determination is so strong on the part of the patient to make away with himself, that it becomes the business of his life to watch for his opportunity. An instance of this kind occurred in the case of a gentleman who was placed under medical care, having attempted to commit suicide. He was watched with the most scrupulous attention; during nine months every means, as far as his attendants knew, was removed from him with which he could injure himself, when he was discovered one morning hanging by the neck from his bedstead, quite dead. How he became possessed of the cord was an enigma, which was afterwards solved by the discovery that he had hoarded every piece of string from the parcels of books that had been sent to him within that period. With these, which he had concealed, he had twisted a rope with which he had accomplished his purpose.

There are epidemics of suicide as well as of fever. When any notorious example is made known in the public papers, it is sure to be followed by cases of a similar kind. In the last century many suicidal outbreaks occurred. In this country, Sydenham informs us, there was an example at Mansfield; and at Versailles, in 1793, a terrible furor of this kind seized the people. In that year, out of the then small population, there were no less than one thousand three hundred suicides. Not many years ago the Hôtel des Invalides was the theatre of one of these extraordinary outbreaks. An invalid hung himself on a cross-bar of this institution, and in the following fortnight five other invalids followed his example on the same crossbar, and the epidemic was only stopped by the governor shutting up the passage. Napoleon, by a masterly stroke, stopped a movement of this kind which threatened to spread among his grenadiers. He issued an order of the day to the following effect:—" The grenadier Groblin has committed suicide from a disappointment in love. He was in other respects a worthy man. This is the second event of the kind that has happened in this corps within a month. The First Consul directs that it shall be notified in the order of the day of the Guard, that a soldier ought to know how to overcome the grief and melancholy of his passions; that there is as much true courage in bearing mental affliction manfully as in remaining unmoved under the fire of a battery. To abandon oneself to grief without resisting, and to kill oneself in order to escape from it, is like abandoning the field of battle before being conquered."

That persons who otherwise would not think of suicide are impelled to it by the notoriety given to cases of the kind is undoubtedly the fact. Sir Charles Bell, the surgeon of the Middlesex Hospital, going into a barber's shop one day to be shaved, told the operator of a case of cut-throat that had just come into the hospital. The man had not succeeded to the extent he desired, said the surgeon, but he might have done so easily had he known how to set about it. The barber seemed eager to be informed where the cut should have been made, and Sir Charles described the anatomy of the neck, and the situation of the great artery. The barber listened, and left the room. Not coming back to complete the shaving operation, Sir Charles went to look for him, and discovered him in a yard behind the house, with his throat cut *secundum artem*. No doubt Sir Charles was very grateful that the experiment had not been tried by the insane barber upon himself.

The Suicidal Act.

In cases where there is an hereditary disposition, suicide will often occur among many members of the same family. We have been informed that all the members of a particular family evinced suicidal tendencies when they arrived at a certain age. There appeared to be no exciting cause in any of these cases. They seemed to be quite well, but at a certain age the propensity was suddenly developed. Whether any of them actually committed suicide we are not informed. In another case, seven members of one family in Paris, afflicted with this hereditary tendency to commit suicide, succumbed to it within thirty or forty years. "Some hanged, some drowned themselves, and others blew out their brains. One of them had invited sixteen persons to dine with him one Sunday. The company collected, the dinner was served, and the guests were at the table. The master of the house was called, but did not answer: he was found hanging in the garret. Scarcely an hour before he was quietly giving orders to the servants and chatting with his friends."

It often occurs that persons put an end to their existence immediately after enjoying themselves in the society of friends. "Can these be said to be insane when they commit suicide? Yes, most undoubtedly," says Esquirol. "Do not monomaniacs appear perfectly sane on all other subjects, till the particular idea is started which forms the burden of their hallucination? A physical pain, an unexpected impression, a moral affection, a recollection, an indiscreet proposition, the perusal of a passage in writing, will occasionally revive the thought and provoke the act of suicide, although the individual the instant before should be in perfect integrity of mind and body;" a very profound explanation this, and worthy of the great physician who made it.

Persons have committed suicide in positions which would have been thought impossible to produce suffocation. There are several cases of this kind reported in a French medical journal. A man was discovered hanging by his pocket-handkerchief, suspended from a rope stretched across a granary. His legs were found bent at a right angle backwards, the knees hanging at the distance of a few inches only from a heap of grain on the floor. It would appear that the man, whilst in the act of hanging, must have held his legs clear of the ground in this extraordinary manner.

Another man hung himself from a grating which was not so high as himself: when found, his legs were stretched out before him, and his hips were within a few inches of the ground. A female suspended herself so low that she was obliged to stretch out her legs, one in advance resting on the heel, the other, behind her body, resting upon the toes. There have been cases known in which death has been produced by the suicide simply leaning with the neck against a tightened cord.

A still more remarkable case was that of a schoolmaster in the neighbourhood of London, who hung himself in such a manner from the banisters by his cravat, that his body was found resting entirely upon the stairs. Such cases as these are often open to medico-legal investigation, as murders may sometimes be very conveniently hidden by contrivances of this character.

The suicide of the Duke de Bourbon, in 1830, was supposed, by some of the witnesses at the inquest, to be only a hidden assassination. However, the fact of the suicide was ultimately established. But men have been executed on suspicion of having murdered others, who were afterwards pronounced to have been veritable suicides. Many men who can swim well have been known to tie their hands and legs together before throwing themselves in the water, lest they should be tempted in the dying agony to strike out and save themselves. A lunatic some years ago, in St. Luke's Hospital, drowned herself in the bath-room of that institution in a most remarkable manner. She managed to secrete the key of the bath-room, and to make up a dummy to represent herself in bed, in order to deceive the nurse of the ward. In the middle of the night she stole downstairs, and was found next morning lying with her face downwards in the shallow water of the bath. She must have deliberately kept herself in the horizontal position in the most determined manner to have effected her purpose.

But determined and deliberate suicides of this kind are characteristic of those suffering from confirmed insanity, whereas, in cases of mere impulsive insanity, the patient often regrets his attempt before it is completed, and is cured, in fact, by the attempt. In the case of Sir Samuel Romilly, the loss of blood, it is suggested, relieved the cerebral congestion which impelled him to make the fatal cut. He bitterly repented of his act immediately it was done, and did all in his power to stop the hœmorrhage.

This fact, which is well supported by others, leads to the conclusion, that whilst the person making an attempt upon his life is undoubtedly insane at the moment, yet that he may be perfectly sane the moment after, provided his life has been spared. The impulse may be likened to that which prompts people to leap from great heights. There are thousands of persons who dare not trust themselves in such positions. The same feeling of dread is very common with respect to razors in certain nervous conditions of the body. It would be absurd to suppose that this fear of an impulse can be looked upon as a symptom of insanity.

There are many cases upon the books of children committing suicide after having heard of some example of the kind in their neighbourhood. It seems almost incredible that the imitative faculty in infants of seven or eight should lead them to such extremities, but it is only another example of the total want of knowledge of the sacredness of life which exists almost as a rule in youth. Not only have children hanged themselves, but their brothers and sisters and young companions. Here, again, insanity has had nothing to do with the act.

We could not, within the limits of this article, attempt to give anything like a collection of the remarkable suicides which have taken place, but we may be allowed to quote a few examples that have been placed on record. We have mentioned a few cases of this kind which apparently have been prompted by a morbid desire to astonish; but an instance was afforded at Fressonville, in Picardy, which has a touch of the grotesque in it worthy of a Frenchman. On a sudden the church-bell was heard to ring at an unusual hour in a very agitated manner. Upon the cause being inquired into, it was found that a man had hanged himself to the clapper, and in the agitation caused by his position, the bell rang in the strange manner that had excited attention. The man happily was not dead: the attempt was certainly a grim effort to ring his own death-knell.* In another case

---

* This, and some other curious examples of suicide, we have quoted from the interesting volume, "The Anatomy of Suicide," by the late Dr. Forbes Winslow.

a woman deliberately broke a hole in the ice, placed her head in it, and held it there until she was drowned.

The most extraordinary example of a deliberate attempt at suicide, combined with publicity, we have heard of was made by an Italian named Matthew Lovat, a shoemaker. Dr. Winslow, quoting fom Dr. Bergierri, who records the case, says: " This man determined to imitate the crucifixion, and for this purpose deliberately set about making a cross and providing all the adjuncts of that terrible scene. He perceived that it would be difficult to nail himself firmly to the cross, and therefore made a net, which he fastened over it, securing it at the bottom of the upright beam, a little below the bracket he had placed for his feet, and at the ends of the two arms. The whole apparatus was tied by two ropes, one from the net, and the other from the place where the beams intersected each other. These ropes were fastened to the bar above the window, and were just sufficiently long to allow the cross to lie horizontally upon the floor of the apartment.

"Having finished these preparations, he next put on his crown of thorns, some of which entered his forehead; then, having stripped himself naked, he girded his loins with a white handkerchief. He then introduced himself into the net, and seating himself on the cross, drove a nail through the palm of his right hand by striking its head upon the floor until the point appeared on the other side. He now placed his feet on the bracket he had prepared for them, and with a mallet drove a nail completely through them both, entering a hole he had previously made to receive it, and fastened them to the wood. He next tied himself to the cross by a piece of cord round his waist, and wounded himself in the side with a knife which he used in his trade. The wound was inflicted two inches below the left hypochondre, towards the internal angle of the abdominal cavity, but did not injure any of the parts which the cavity contains. Several scratches were observed upon his breast, which appeared to have been done by the knife in probing for a place which should present no obstruction. The knife, according to Lovat, represented the spear of the Passion.

" All this he accomplished in the interior of his apartment; but it was now necessary to show himself in public. To accomplish this, he had placed the foot of the cross upon the window-sill, which was very low, and by pressing his fingers against the floor he gradually drew himself forward, until, the foot of the cross overbalancing the head, the whole machine tilted out of the window and hung by the two ropes which were fastened to the beam. He then, by way of finishing, nailed his right hand to the arm of the cross, but could not succeed in fixing the left, although the nail by which it was to have been fixed was driven through it, and half of it came out on the other side.

" This took place at eight o'clock in the morning. Some persons by whom he was perceived ran upstairs, disengaged him from the cross, and put him to bed. By medical care his wounds ultimately healed, and the poor man recovered his mental condition, but he was, however, ever afterwards morose and singular."

We have said that November is debited with the greatest number of suicides : this is an invention of the French wits, who love to rail at our climate. In fact, November seems to be avoided by the suicide; it is too uncomfortable for the attempt. The months of March, June, and July record the greatest number of suicides, but females appear to prefer September, November, and January. The diversity of the sexes with respect to the season chosen for suicide is remarkable. The males are always in excess of the females with regard to this dreadful act. Possibly the reason of this is the wider choice of means of self-destruction possessed by the men. The gun and the razor are generally at hand, whereas water, the usual means by which the female makes her exit, has to be sought sometimes at a long distance. Women but rarely cut their throats. With men in early life hanging is the method preferred, fire-arms in the more vigorous, and hanging again in advanced life. This difference is clearly owing to the fact that fire-arms are but rarely used by a man except in middle age. We may add that poisoning is a favourite method of suicide with young persons, especially females, and charcoal also; but that is a French invention, which has not yet touched the rougher instincts of the Briton.

When the act of suicide has been caused by some overwhelming misfortune, it certainly need not be ascribed to insanity; there are rare occasions when the individual may be excused for putting an end to a life which circumstances have made intolerable to him; but causes of this overpowering nature are but rare, compared with those that are purely impulsive. Even the impulse differs vastly in its intensity and persistency. Many an individual has attempted his life, and whilst in the act has regretted it, and struggled to save himself, whilst others have contemplated its accomplishment as a fate which they cannot possibly resist.

It is unjust and unphilosophical to speak of both as being equally insane. The mere impulse to commit suicide is very often of a temporary nature, and incited by the presence of some immediate means of perpetrating it, which an individual so liable to be tempted, if he is sufficiently sane, carefully avoids. There are, possibly, few persons of an impulsive nature who have passed near great heights, dark streams, or have crossed the Clifton Suspension Bridge, for instance, with a view of the yawning gulf beneath them, without experiencing a latent dread lest they should be tempted to jump over. There are others, again, who go to such places for the express purpose of committing suicide. The power of the will to conquer the impulse, or to repress the contemplation of putting it in action, is the sole difference between the healthy and the insane mind. The impulse itself appears to be one of those inexplicable mental phenomena which haunt those who dwell on the Borderland of Insanity.

After resisting, on many occasions, a time may come to the feeble-minded when the impulse conquers the resistance, and the suicidal act is the result. Only the individual himself is aware of the terrible struggles that, under the calmest exterior, sometimes agitate him. Happy he whose nervous temperament does not render him liable to these frightful conflicts with his better reason!

THE END.

WYMAN AND SONS, PRINTERS, GREAT QUEEN STREET, LONDON, W.C.

www.ingramcontent.com/pod-product-compliance
Lightning Source LLC
Chambersburg PA
CBHW030749230426
43667CB00007B/900